New Directions in Philosophy and Cognitive Science

Series Editors: **Richard Menary**, Macquarie University and **John Sutton**, Macquarie University

This series brings together work that takes cognitive science in new directions. Hitherto, philosophical reflection on cognitive science – or perhaps better, philosophical contribution to the interdisciplinary field that is cognitive science – has for the most part come from philosophers with a commitment to a representationalist model of the mind.

However, as cognitive science continues to make advances, especially in its neuroscience and robotics aspects, there is growing discontent with the representationalism of traditional philosophical interpretations of cognition. Cognitive scientists and philosophers have turned to a variety of sources – phenomenology and dynamic systems theory foremost among them to date – to rethink cognition as the direction of the action of an embodied and affectively attuned organism embedded in its social world, a stance that sees representation as only one tool of cognition, and a derived one at that.

To foster this growing interest in rethinking traditional philosophical notions of cognition – using phenomenology, dynamic systems theory, and perhaps other approaches yet to be identified – we dedicate this series to "New Directions in Philosophy and Cognitive Science."

Titles include:

Miranda Anderson
THE RENAISSANCE EXTENDED MIND

Robyn Bluhm, Anne Jaap Jacobson and Heidi Maibom (*editors*)
NEUROFEMINISM
Issues at the Intersection of Feminist Theory and Cognitive

Jesse Butler
RETHINKING INTROSPECTION
A Pluralist Approach to the First-Person Perspective

Massimiliano Cappuccio and Tom Froese (*editors*)
ENACTIVE COGNITION AT THE EDGE OF SENSE-MAKING
Making Sense of Non-sense

Maxime Doyon & Thiemo Breyer (*editors*)
NORMATIVITY IN PERCEPTION

Matt Hayler
CHALLENGING THE PHENOMENA OF TECHNOLOGY

Anne Jaap Jacobson
KEEPING THE WORLD IN MIND
Mental Representations and the Sciences of the Mind

Julian Kiverstein & Michael Wheeler (*editors*)
HEIDEGGER AND COGNITIVE SCIENCE

Michelle Maiese
EMBODIMENT, EMOTION, AND COGNITION

Richard Menary
COGNITIVE INTEGRATION
Mind and Cognition Unbounded

Zdravko Radman (*editor*)
KNOWING WITHOUT THINKING
Mind, Action, Cognition and the Phenomenon of the Background

Matthew Ratcliffe
RETHINKING COMMONSENSE PSYCHOLOGY
A Critique of Folk Psychology, Theory of Mind and Stimulation

Jay Schulkin (*editor*)
ACTION, PERCEPTION AND THE BRAIN

Tibor Solymosi and John R. Shook (*editors*)
NEUROSCIENCE, NEUROPHILOSOPHY AND PRAGMATISM
Brains at Work with the World

Rex Welshon
NIETZSCHE'S DYNAMIC METAPSYCHOLOGY

New Directions in Philosophy and Cognitive Science
Series Standing Order ISBN 978–0–230–54935–7 Hardback
978–0–230–54936–4 Paperback
(*outside North America only*)

You can receive future titles in this series as they are published by placing a standing order. Please contact your bookseller or, in case of difficulty, write to us at the address below with your name and address, the title of the series, and one of the ISBNs quoted above.

Customer Services Department, Macmillan Distribution Ltd, Houndmills, Basingstoke, Hampshire RG21 6XS, England

---

A Neurophenomenology of Awe and Wonder

Towards a Non-Reductionist Cognitive Science

Shaun Gallagher
University of Memphis, USA and University of Wollongong, Australia

Lauren Reinerman-Jones
University of Central Florida, USA

Bruce Janz
University of Central Florida, USA

Patricia Bockelman
University of Central Florida, USA

Jörg Trempler
Humboldt University, Germany

palgrave
macmillan

First published 2015 by
PALGRAVE MACMILLAN

Palgrave Macmillan in the UK is an imprint of Macmillan Publishers Limited, registered in England, company number 785998, of Houndmills, Basingstoke, Hampshire RG21 6XS.

Palgrave Macmillan in the US is a division of St Martin's Press LLC, 175 Fifth Avenue, New York, NY 10010.

Palgrave Macmillan is the global academic imprint of the above companies and has companies and representatives throughout the world.

Palgrave® and Macmillan® are registered trademarks in the United States, the United Kingdom, Europe, and other countries.

ISBN: 978–1–137–49604–1

This book is printed on paper suitable for recycling and made from fully managed and sustained forest sources. Logging, pulping, and manufacturing processes are expected to conform to the environmental regulations of the country of origin.

A catalogue record for this book is available from the British Library.

A catalog record for this book is available from the Library of Congress.

Contents

List of Illustrations

Figures

Tables

Acknowledgments

This book is the result of a large research project that ran officially for two years (2011–2013), and unofficially is still ongoing. A large number of people deserve our thanks for their input and support of this project. First, the complete team of research collaborators who worked on it at various times at the Institute for Simulation and Training at the University of Central Florida: Steve Fiore (Philosophy and Cognitive Science Lab), Stephanie Lackey (Active Lab), Eileen Smith (Media Convergence Lab), Brandon Sollins (Active Lab), Mike Carney (Media Convergence Lab). Also, Prof. Garrett Riggs (College of Medicine: Neurology, UCF), and, at the Forscherkolleg Bildakt und Verkörperung, Humboldt University, Berlin: Horst Bredekamp, Joerg Fingerhut, and Matthias Bruhn.

We also want to thank our two external consultants on the project: Jonathan Cole (Neuroscience, University of Bournsmouth), who has worked with NASA on a number of projects, and Jeffrey Williams (NASA astronaut), who so far has spent 362 days in space, is fourth on the U.S. list of long-duration space travelers, and is heading back to space in 2016.

We greatly appreciate comments we received on this project from participants at the workshop on Space, Science, and Spirituality at the Forscherkolleg Bildakt und Verkörperung, Humboldt University (July 2012), and the Conference on Awe and Wonder at the University of Central Florida in September 2013, including Jesse Prinz and Michelle Shiota. In addition, Piet Hut, at Princeton University, Patrick McGivern, at the University of Wollongong, and Dan Zahavi, at the University of Copenhagen, offered some helpful comments on the penultimate draft of this book.

The largest official thanks goes to The John Templeton Foundation which funded most of our research for this project. A variety of institutions also helped to support our research: the Institute for Simulation and Training at the University of Central Florida, the Philosophy Departments at the University of Memphis and the University of Central Florida, the Kolleg-Forschergruppe Bildakt und Verkörperung at the Humboldt University, and the Philosophy Department of the Faculty of Law, Humanities, and the Arts at the University of Wollongong. Shaun Gallagher also thanks the Humboldt Foundation's Anneliese Maier

Research Award, for ongoing support that allowed him to continue to pursue this project.

Some of the material in this book includes revised segments from the following research papers. We thank the journals for permission to use this material:

1. Bockelman-Morrow, P., Reinerman-Jones, L. and Gallagher, S. 2013. Methodological lessons in neurophenomenology: Review of a baseline study and recommendations for research approaches. *Frontiers in Human Neuroscience*. doi: 10.3389/fnhum.2013.00608.
2. Gallagher, S., Reinerman-Jones, L., Sollins, B. and Janz, B. 2014. Using a simulated environment to investigate experiences reported during space travel. *Theoretical Issues in Ergonomic Sciences* 15(4): 376–394.
3. Reinerman-Jones, L., Sollins, B., Gallagher, S. and Janz, B. 2013. Neurophenomenology: An integrated approach to exploring awe and wonder. *South African Journal of Philosophy* 32(4): 295–309.

1
Liftoff: Towards an Exploration of Subjective Experience

At one View ten thousand sparkling Orbs survey,
Innumerable Worlds and dazzling Springs of Light.
O the vast Prospect! O the charming Sight!
How full of Wonder, and Delight!
How mean, how little, does our Globe appear!
This object of our Envy, Toil and Care,
Is hardly seen amidst the Crowd above;
There, like some shining Point, does scarce distinguish'd move.
(Lady Mary Chudleigh, 1703)

10 By way of introduction ... a countdown

Near the ruins of the ancient city of Miletus, you can still walk out into an open field at night and gaze at an extremely rich array of stars. According to a famous legend, in the sixth century BCE, Thales of Miletus, one of the first philosophers to appeal to naturalistic explanations, walking across a field and gazing at the stars, found the heavens so wondrous, or was so lost in his astronomical calculations, that he walked directly into a well. Wonder has a double meaning nicely captured in the uncertainty of Thales' mental state. Was he so awestruck by the starry vista that he was caught up in the reflective emotion of *wonder*, or was he so busy just *wondering*, intellectually, how the heavens worked? The two senses of the term meet in the claim that wonder is the beginning of philosophy. The first sense is closely tied to the feeling of awe; the second to the feeling of curiosity.

Aristotle, another Greek philosopher, argued that all humans desire to know – to pursue knowledge. Humans also desire experiences of awe and wonder. Some empirical evidence for this can be found every evening in desert places like Sedona, Arizona where small crowds gather on various hills and vortexes to witness sunsets that generate awe-inspiring and wondrous light effects on the nearby mesas and rock formations. Many people live near, or travel to the sea or mountains to have similar experiences.

Experiencing views of the stars from Earth, and views of various features of Earth from various places on the Earth's surface can be sources of awe, wonder, curiosity and even humility. The central topic of this book involves experiencing views from a completely different perspective – looking at Earth and the surrounding universe from a position in outer space, that is, outside of Earth's atmosphere.

9 To infinity and beyond

A significant number of astronauts, cosmonauts, and other space travelers[1] have reported experiences that are deeply aesthetic, spiritual, or sometimes religious as they have orbited the Earth in the Space Shuttle or on the International Space Station (ISS). Some of these space travelers have kept journals that describe these experiences during space flight. In many cases, they have reflected on these experiences after their return to Earth. The astronauts reporting such experiences come from different backgrounds – many are engineers and hard-nosed scientists; some are atheists; and some have religious backgrounds. Some have been led by these experiences to become more spiritually sensitive or more attuned ecologically or ethically after their return to Earth. For a few of the astronauts, these experiences have been life transforming.

On the face of it, awe and wonder are experiences that transcend religion, culture, politics, and just about every other "contextualizing" feature of human existence that one could imagine. Do we not think that almost everyone experiences such phenomena at various points in their lives under various circumstances? It seems that senses of awe and wonder come as close to human universals as could be imagined, and if we look at long-standing philosophical discussions of such things, the result of that universality has been that we have gone outside of ourselves to explain or understand our senses of awe and wonder. Such experiences must be gifts from a divine being, or from the universe itself, or from our evolutionary past as a reaction to the unknown (the "survival of primitive thought," as William James called religion). It must, in short, be transcendental in some way.

[1] Hereafter we refer to "astronauts" without distinguishing between cosmonauts or other non-astronaut space travelers.

And yet, that move has, until recently, taken us away from considering awe and wonder as phenomena worth understanding in their own right. In the nineteenth century, however, during the rise of the scientific study of religion, theorists did try to categorize "spiritual" experience (by which they meant any experience beyond a sensory awareness of the world). The goal of such categorization was to show that the spiritual was not, after all, transcendent, but immanent. The move to regard as immanent any sensibility other than the mundane rendered all human experience available to scientific investigation. However, this was followed by a backlash among those who felt that to study the awe or wonder experience fundamentally diminished it. From this perspective, it was a classic case of an observer effect in which studying something fundamentally compromises it. And yet, those who wanted to preserve awe *qua* awe and wonder *qua* wonder sometimes themselves fell into the same trap in which those experiences were explained by reference to a specific religious tradition or, later, psychotropic causes. The experiences of awe and wonder were still there, but they were rendered understandable through explanatory (quasi-causal) structures.

8 Scientific frontiers

This book is about the first scientific study of these experiences in the context of space travel.[2] The aim of this study was to explore what traditionally might be called the *inner space* of experience, while traveling in *outer space*. Using this vocabulary of inner *versus* outer, however, is not the best way to put it, even if it connects with considerations that go back centuries. The study of the experience of those who have travelled to space avoids the temptation to either reduce awe and wonder to mundane experience or explain it completely by reference to some internal processes caused by external stimuli. There are, of course, stimuli present – the sunrises that occur every 90 minutes while in orbit, the deep blackness of space, the land formations that can be viewed while 230 miles above the earth in the International Space Station. However, these are not causal in the sense that divine agency or LSD might be causal. They are contextual.

---

[2] The research was conducted by the co-authors of this book, who were part of an interdisciplinary research team based at the University of Central Florida's Institute for Simulation and Training, but included researchers from the University of Memphis and the Humboldt University in Berlin. The research was funded by a grant from the Templeton Foundation.

For our study, we were interested specifically in replicating the consciousness involved in the experiences of awe and wonder had by astronauts during space travel and in examining such experiences from the perspective of the interdisciplinary cognitive sciences. We gathered together a research team of psychologists, neuroscientists, and philosophers. Also, to do this study without incurring the large cost of sending people into space, we were joined by engineers and experts in the art of simulation who were able to create virtual environments that we used as scientific test beds. For the majority of us, this was the most interdisciplinary study we have been involved in. The interdisciplinary organization of the study itself was a challenge that broadened our understanding of the nature of scientific investigation and motivated, in effect, a more intellectual and collectively shared type of wonder. This book, then, has two aims: (1) to explain the awe and wonder experienced during the adventure of space travel, and (2) to give an account of the scientific adventure of studying and wondering about such experiences.

7 Starlogs of experience

The subjective experiences of awe and wonder that we wanted to study have been well documented in journals written by astronauts during their time in space. After returning to Earth, they also wrote reflections and talked about their experiences in interviews. To initiate our scientific project, we gathered these various descriptions together and began to analyze them in two different ways. We first did a hermeneutical analysis. Hermeneutics is a branch of philosophy concerned with the interpretation of texts. It involves a careful, self-conscious analysis of the meaning of texts that keeps in mind the historical or biographical background of the authors, their intentions in writing the text, the audience they intended to reach, and the specific vocabulary they had available. Additionally, it takes into account the intentions and practical interests of the interpreter. Specifically, in this analysis, our focus was on descriptions of the experiences themselves, and we were concerned to define in very precise terms, the different forms that such experiences took. Second, we did a computer-based syntactical analysis. We looked at various patterns of syntactical structure in the language used by the astronauts to express their experiences. For example, by this method we found that the descriptions contained in the in-space journals were much more concrete (or less abstract) than the descriptions provided in later reflections and interviews.

Here are some sample descriptions that we analyzed (the details of these analyses are provided in Chapter 2).

> There is a clarity, a brilliance to space that simply doesn't exist on Earth. And nowhere else can you realize so fully the majesty of our Earth and be so awed at the thought that it's only one of untold thousands of planets. (Gus Grissom, USA)
>
> From these windows, the Earth is so obviously floating in an endless void, and the feeling that washes over you is the sense of scale of the universe. The feeling I got was one of recognition that "we are living on a such a tiny island in a vast ocean." (Greg Chamitoff, Canada)
>
> One thing is just the sheer wonder of looking down at the Earth. It is very, very beautiful... particularly when you see it interface on the edge with space. There you suddenly get the feeling that, hey, this is just one small planet which is lost in the middle of space.... a very important feeling about the fact that we're just drifting through an immense universe.... you become a little more conscious about the fact that we shouldn't be doing silly things on Earth like fighting and killing each other. (Marc Garneau, Canada)
>
> I frequently recalled what King David had written thousands of years ago in Psalm 19: "the heavens declare the glory of God, and the firmament showeth his handiwork." That order, that creation was very apparent to me as I looked back on this beautiful planet that looked so fragile at the same time. (Bill Nelson, USA)
>
> The atmosphere is very small, very thin compared to the universe that surrounds the Earth. It made me feel that the Earth was very fragile and unprotected... and it worried me. It worried me. (Alexey Leonov, Russia)
>
> In the first glance as I looked at the Earth and saw the blue color of the Earth, I said "Suppana Allah" – something that in Arabic would mean "God is great." (Prince Al Saud, Saudi Arabia)

Most of the descriptions are based on the astronauts' views from windows in either the Space Shuttle or the ISS, or, less frequently, from what they could see during space walks. You can identify various themes in these expressions, ranging from aesthetic appreciation, to ecological and ethical concerns, to religious feelings. This is a small sample. We analyzed texts from 45 astronauts, including 17 in-flight journals, and 34 post-flight interviews and reflections – a total of

23,000 words. As we indicated, our focus was on the phenomenology, that is, the experiential aspects of these descriptions. For example, we frequently found what we came to call "scale effects" – changes in perspective concerning relative size – feelings of smallness in contrast to the vastness of the universe. Greg Chamitoff expresses this metaphorically: "we are living on a such a tiny island in a vast ocean." Marc Garneau puts it in direct terms of his experience: "you suddenly get the feeling that, hey, this is just one small planet which is lost in the middle of space…a very important feeling about the fact that we're just drifting through an immense universe." This is one of the very specific kinds of experiences that we came to consider as an instance in the broader category of awe.

6 Clearing our heads

In a hermeneutical analysis, the interpreter needs to become aware of his or her own biases. One may not be able to escape all such biases, but it is important to identify them and to lay them out on the table for all to see. One such bias is very basic: When we went looking for experiences of awe and wonder, we already had some conception of what we were looking for. We self-consciously decided on some working definitions. Although one of us (Bruce Janz) is well versed in the history of mysticism, we took a more pragmatic (less historical) route to defining awe and wonder, treating the definitions as tentative, preliminary, provisional, and open to revision. Still, these definitions guided our reading of the texts. However, we also found certain things of interest that did not fall into these categories, and so widened our scope as we did our analysis. Here are the working definitions with which we started.

- **Awe:** a direct and initial experience or feeling when faced with something amazing, incomprehensible, or sublime
- **Wonder**: a reflective experience motivated when one is unable to put things into a familiar conceptual framework – leading to open questions rather than conclusions

Awe hits you more immediately at the first-order level of experience; wonder is more reflective or second-order. One can think that perhaps an immediate experience of awe motivates a more reflective experience of wonder. We have more to say about these definitions, and about what we found in the astronauts' texts in the next chapter.

5 Consulting the phenomenological map

Although we had some idea of what we were looking for, it was not at all clear how we would be able to study such things. We started by asking the following questions.

1. Can we characterize these experiences in their own terms as experiences?
2. Can we replicate them in a lab setting?
3. Can we measure the physiological and neurophysiological correlates of these experiences?
4. Can we say something about how a person's background might play into the experiences?

The first question is about *phenomenology*. The term, "phenomenology" has two meanings in contemporary philosophy. In one sense (used primarily in philosophy of mind and cognitive science), it just means the consciousness that someone experiences – more specifically, the phenomenal aspect of consciousness, the qualitative feel, or what Thomas Nagel (1974) has called, the "what it is like" to experience a sensation or to perceive some object. For example, there is something it is like for you to experience a pain in your foot. This particular experience is different from what it is like to taste chocolate. Likewise, we can think that there is something it is like to experience awe – and what we want to know is precisely what that experience is like.

The second meaning of the term "phenomenology" refers to a particular philosophical approach to studying experience. This approach originated with the German philosopher, Edmund Husserl (1859–1938), and was further developed in the philosophical works of Heidegger, Sartre, and Merleau-Ponty, among others. This phenomenological approach provides a method for answering the first question, since phenomenology is just an attempt to characterize experiences in one's own terms. The idea that we can use a phenomenological approach as part of a scientific explanation of experience, however, is controversial. We map out this controversy briefly here and return to it in Chapter 4.

Science has traditionally been considered to be third-person and reductionistic. "Third-person" means that scientists attempt to take a neutral (or objective) observational stance on what they study. A chemist, for example, is not personally involved with her chemicals, and she is supposed to be objective in her approach to explaining how they work. She shouldn't allow her own preferences to enter into her analysis, and scientific method introduces controls in order to make her results as objective as possible. Any other scientist, following the same procedures, should be able to replicate her results. In practice, of course, the research questions that any particular scientist asks may be biased by her own theoretical approach or sometimes by what funding agencies are most interested in. Once the question is defined, however, the scientist is supposed to pursue an answer guided by methods that can be replicated, at least to the point of producing the experimental data, and prior to any interpretation of what the data mean. Once interpretation starts, so do the debates. Although some of these debates have been settled by appealing to particular paradigmatic theories, much of science is still open to debate, which is what makes it interesting.

Science also tends to be reductionistic. The chemist wants to show how certain higher-order properties of a certain material can be explained by a lower-order organization of molecules, and on an even lower scale, precisely how such molecules work at an atomic level, and, in some cases at a quantum level involving the behavior of sub-atomic particles. Reductionism is often thought of in terms of translating from a higher-order theory or vocabulary to a more basic one. Some sciences may focus on water as important for life, as part of weather systems that cause draught or floods, and so on, but ultimately all science agrees that, at the most basic level, water is defined as H_2O. This kind of agreement about the nature of the object under discussion is what reductionistic science aims at. It's clearly a good thing to aim for since it tells us something important, and it leads to a consensus about the object of study.

Both of these tendencies in science are carried over into the scientific study of human behavior and experience. Behaviorism, in the first half of the twentieth century, famously replaced introspection as the preferred scientific method because introspection was thought to be too subjective and unverifiable. A person may be able to report on his experience, but there is no way to verify that his report is veridical, or even that the experience he calls pain, for example, is precisely the same experience a different subject describes as pain. Behaviorism limited data to observable behavior – responses that could be measured in third-person, objective terms. Although cognitive science turned its attention to the inner workings of the cognitive system, it still followed the behaviorist model in terms of measuring what could be observed experimentally. The famous Turing Test, devised by Alan Turing (1950), remained behaviorist in that it judged a system's intelligence on the basis of its observable responses. The computational models developed for artificial intelligence allowed scientists to build and test systems with a complete knowledge of their mechanical design. And neuroscience has developed sophisticated methods for observing (or at least inferring) brain behavior.

Likewise, cognitive neuroscience aims for reductionist explanations. We explain higher-order cognitive processes ultimately in terms of neuronal processes, and some would argue that the best explanations of cognition are found at the level of molecular neuroscience (Bickle 2003), or even at the quantum level (Penrose 1999). Philosophers, from Dan Dennett (1991) to Patricia Churchland (2011; 2013), argue that explanations in terms of consciousness or conscious intentions (what Dennett [1989] calls the "intentional stance") are, at best, placeholders for stricter scientific explanations in terms of computational design or neuronal physics.

In contrast to such third-person, reductionist explanations, the phenomenological approach is first-person and non-reductionist. Some cognitive scientists tend to regard phenomenology in this sense as similar to introspection, and therefore as subjective and unreliable (Dennett 1991). When the subject matter is consciousness itself, however, that is, when what we are trying to explain is a subject's experience, things are more complicated. Consciousness is itself first-personal – that is, it is always something that occurs in first-person perspective. As William James (1890) put it, consciousness is always personal. It involves a subject or person experiencing something from an egocentric or body-centered perspective; and, as such, the experience is subjective. This motivates two questions. First, can we study something that is subjective and first-person (such as the feeling of pain, or experiences of awe and wonder) in a purely objective third-person way without losing something in the process? Do we get a complete explanation of pain, for example, if we provide an account purely in terms of neuronal activation of nociceptive C-fibers? Or is there something more to it? Second, in the study of consciousness or cognition, even if we want to give an explanation in terms of the subpersonal (neuronal) processes involved (the *explanans*), we also need to know precisely what is it that we are trying to explain in any particular instance (the *explanandum*). If we are trying to explain experience of one sort or another, for example, then, as the phenomenologists argue, we need to have a good description of the *explanandum* that we are aiming to explain. It's not enough for the neuroscientist to say that C-fibers are firing; she also has to say precisely what this explains. Indeed, it would be methodologically impossible to engage in neuroscience without referring in some way to behavior or experience, and since the latter is first-personal, it requires that the account be first-personal, or include the first-personal perspective in some respect.

Phenomenologists thus argue that, if the subject matter is experience (consciousness), we need a good, and methodologically controlled, description of the particular experience that we are studying, and that it would be unscientific to pretend that we could understand such phenomena in a strictly third-person way. This involves a vocabulary that includes higher-order cognitive and experiential terms and not just the terminology of neurons and molecules.

4 Probing experiences

The third question on our list, however, asks whether we can measure the physiological and neurophysiological correlates of the experiences of awe and wonder. So, we are not interested only in the phenomenology of these experiences, but we also want to know what is happening on a subpersonal, neuronal level. In this regard, the method that seems best able to deal with both phenomenology and neurological processes is one proposed by the late neurobiologist Francisco Varela (1996). Varela proposed an approach that delivered on the idea that we can take both first-person data seriously in a correlational analysis with third-person data, without reducing one to the other. Borrowing from the phenomenological approach outlined by Husserl, Varela defined "neurophenomenology" as a method, or combination of methods, that involves the training of subjects in phenomenological method, and then using these subjects as participants in empirical experiments.

Husserl's phenomenological method, as adapted by Varela, involves three steps.

1. Suspending beliefs or theories about experience;
2. Gaining intimacy with the domain of investigation;
3. Offering descriptions and using intersubjective validations.

The first two steps have technical names. Husserl called the first step the *epoché*, a Greek term usually translated as "bracketing" in this context. To suspend or bracket one's beliefs is not to enter into a skeptical doubt about those beliefs, but simply to set them aside. The point is to direct one's attention to the experiences *as such* rather than to one's opinions or beliefs about what the experience means or how it might be caused. Setting aside one's theories includes setting aside any scientific or metaphysical theories about the experience in order to get to the experience in its own terms. If I am to give a strict phenomenological description of the pain that I feel, the Gate Control Theory of Pain (Melzack and Wall 1967), which may be scientifically correct, is not part of my description since I'm not conscious of anything like a gate control mechanism in my spinal cord.

The second step is usually referred to as the "phenomenological reduction," where the term "reduction" is understood in light of its Latin roots in the verb, *redūcere,* meaning "to bring back." The idea is to bring one's attention back to the experience itself. In contrast to the *epoché,* which excludes beliefs and theories, this is a more positive step where we develop ways to express what we experience. Engaging in this involves some practice of attending to various aspects of experience without reifying them or turning them into objects that one is simply observing in reflection. In this regard, it's more about *how* I, as subject, am experiencing something rather than about the *what* I am experiencing. Or, again, the task is to say *what it is like* to experience (e.g., seeing an apple tree) rather than what *it* (the apple tree) is like.

The final step has the practical effects of verifying the phenomenological description and helping to develop a shared vocabulary for such experiences. Nothing guarantees that we will all experience the world in precisely the same way. But communicating with others about one's experience, or comparing many different descriptions of a particular experience, can help to elucidate differences and similarities. The phenomenologist aims to discover the invariant features of experience or to discover the basic structures that seem to apply to all experiences of a specific type. The *neurophenomenologist* may be more interested in how precisely similar experiences correlate with neurological processes (Gallagher 2003; Gallagher and Brøsted Sørensen 2006; Lutz 2002; Varela 1996).

In some of Varela's studies, the subjects were trained in the phenomenological method. Not all subjects, however, can undergo phenomenological training. Accordingly, working with epileptic patients in Paris, the Varela group developed a phenomenological interview technique that can substitute for training experimental subjects to participate in neurophenomenological experiments (Petitmengin 2006; Le Van Quyen and Petitmengin 2002). Subjects who are not trained in phenomenological method participate in experiments and are then interviewed by trained phenomenological interviewers. In this process, subjects are "led back" to their experiences and encouraged to explicate them in their own terms (enacting a phenomenological reduction). The interviewers keep the subject's attention focused on the experiential aspects rather than on their opinions, beliefs, or theories about their experience (enacting the *epoché*). Across a sufficient number of subjects one can also attain intersubjective validations in this manner.

In Chapter 4, we'll detail how the neurophenomenological method, combining phenomenology with neuroscientific techniques using EEG, has been used in experiments conducted by the Varela group in Paris (e.g., Lutz et al. 2002), and how we have been using a variation of this method, combining EEG and Functional Near-Infrared Spectroscopy (fNIR) to measure what happens in the brain while subjects have experiences of awe and wonder in simulated space travel. Our study is a neurophenomenological study that also incorporates a phenomenological interview method.

3 Rolling out the space craft

The second question on our list posed a serious challenge. Would we be able to place people in an environment that would elicit just the experiences we were looking for and thereby replicate the astronauts' experiences? There were a number of limitations involved in our solution. First, given certain financial limitations, we were unable to send our subjects (over 100 of them) into space. To build a space-like environment on Earth, however, we recruited computer and simulation engineers, artists, and art historians to design two different kinds of simulations. We conducted two experiments: the first using a mixed reality environment (which we called the Virtual Space Lab [VSL]), and the second using an entirely virtual environment. Mixed reality combines real objects and virtual reality. The VSL involved building a simulation of an ISS workstation. We minimized the instrumentation and made the windows more prominent, since we were interested primarily in what subjects would experience as they looked out of the windows (Figure 1.1).

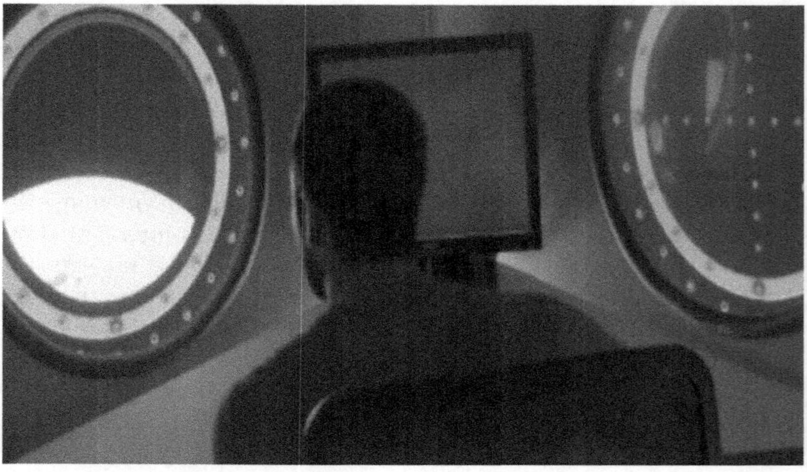

Figure 1.1 Interior of the VSL

In addition to the physical structure, we had to design a dynamic simulation of space scenes that astronauts would have viewed during their real space flights. One of the first tasks in this process was to select the images to be used as the basis for the virtual scenes. To help us select the right images, we consulted with researchers at *The Collegium for the Advanced Study of Picture Act and Embodiment* at Humboldt University in Berlin. This is a group of art historians and philosophers, organized by Horst Bredekamp and the late John Krois, whose research projects include the study of the historical use of images in science, engineering, and other contexts. Bredekamp's concept of picture- or image-act (*Bildakt*) signifies the study of how people use images and what images do to (or for) us in specific contexts.

Members of the *Bildakt* group, including Jörg Trempler, examined thousands of images and conducted an analysis of ones appropriate for our study, taken from the NASA Image Data Base (http://www.nasa.gov/multimedia/imagegallery/index.html). The NASA database is classified according to four major categories (item, location, person, time). These categories do not differentiate between media, contexts, or individuals, so that specific keywords like "earth" (23,491 entries) or "sun" (16,092 entries – accessed in January 2012) also return a large number of diagrams, models, artistic views (drawings, paintings), documentary photographs of laboratories, staff etc. Less than 10% of all images in the gallery were considered relevant to our project. To guide us, the Bildakt group defined a set of criteria for image selection for the VSL (see Chapter 3).

The VSL presented two different visuals to each subject: one involving a view of the Earth from near-Earth orbit and the other involving deep space. A different simulated environment was used in the second experiment. For reasons to be explained in Chapters 3 and 4, we needed to exclude various complicating factors found in the VSL design and to focus on a purely visual stimulus. Accordingly, we moved the second experiment to a VR cave and immersed subjects within a large visual of near-Earth orbit. Details of this experimental design will be made clear in Chapter 5.

2 Down to earth

There were clearly some limitations involved in both experiments. First, in contrast to the situation of astronauts in space, we were not able to replicate weightlessness. Although there are some ways to simulate microgravity, none of them were consistent with the idea of presenting visual simulations of space during the experiments. Although we could not control for the fact that vestibular and proprioceptive systems responsible for balance and body position sense are intermodally tied to the visual system and are dramatically affected (or readjusted) during weightlessness (Harris et al. 2010), we can point out that astronauts (and their vestibular and proprioceptive systems) quickly adapt to microgravity and there is little mention of the challenges of moving around in a weightless state after a few days in space (see Chapter 2 for further discussion). From the astronauts' journals, we can definitively say that the descriptions of awe and wonder experiences are closely tied to seeing the visual environment outside of the windows and seemingly not to the experience of weightlessness. In addition, like the astronauts in some instances, the subjects were strapped into a workstation chair and therefore would not be floating even if they were in a microgravity environment.

Not only were we weighed down by Earth's gravity, we were also indoors, on Earth, and not moving in a fast orbit; and every one of our subjects knew this. There is an important difference between knowing that one is risking one's life, from launch time to return landing; knowing that one is floating in the relative nothingness of space around the earth; and knowing that one is sitting safely in a simulation. Our subjects were not astronauts, did not have the training of astronauts, and were not taking any real risks. Granted all of this, and despite it, we did get some very interesting results and were able to replicate the experiences of awe and wonder. Whether these experiences were as intense as they were for astronauts remains unclear. Also, whether the experiences undergone in the experiments motivated any profound changes in our subjects' lives is an ongoing research question.

1 A final check

To answer three of our four questions, we had to draw on the expertise of philosophers (experts in hermeneutics and phenomenology), neuroscientists, art historians, and simulation engineers. The experimental design was the responsibility of two psychologists and an expert in simulation and modeling. Our fourth question came back again to the psychologists. Can we say something about how a person's background might play into their experiences in the simulations? As we just mentioned, the participants in our experiment were quite different from astronauts in terms of their training and, likely in terms of their amenability to risk taking, and along several other measures. But in many other ways, including their cultural and religious backgrounds, the astronauts who reported experiences of awe and wonder differed from each other. We wanted to know how differences in personality, intelligence measures, cognitive factors, and cultural and religious practices might enter into our subjects' capacities for the experience of awe and wonder. To find out such things, we administered a battery of questionnaires designed to provide psychometric data and data on the cultural practices of the participants. For example, we were able to gain measurements of our participants' personality traits and emotional intelligence, their tolerance for ambiguity, their openness to absorbing self-altering experiences across a number of scales, and their attitudes and practices relevant to religiosity/spirituality, as well as levels of individual and collective self-concepts.

The questionnaires provided a lot of information on the participants in the experiments – some partial indication of their personality traits, their cognitive and emotional life, and their cultural practices. Together with the first-person data provided by the phenomenological interviews, and some further verification tests – for example, an additional experiment-specific survey designed to provide quantitative data on first-person experience – the battery of questionnaires allowed us to go beyond the type of narrow causal account that one often finds in cognitive neuroscience (e.g., these neurons cause this experience – end of story) and to incorporate a more well-rounded picture of how these broader aspects factor into the experiences of awe and wonder.

This "large" methodology, which combines information from phenomenology, psychology, and neuroscience in a tightly controlled environmental setting, we propose, is one way to do a non-reductionist cognitive science and to study consciousness more generally (see Chapter 8). Not all causal, or constitutive factors of experience are to be found simply in the brain. Cognition is embodied and situated in rich social and cultural environments. A non-reductionist cognitive science attempts to grasp as much of these non-neural factors as possible without ignoring the important role of brain processes. To understand what the brain is doing, however, we need the broader picture that involves experiential, embodied, socially and culturally situated factors that contribute to make each person's experience what it is.

0 Ignition

Our analysis gets off the ground in Chapter 2 where we provide details about the textual analysis of the astronauts' in-flight journals and post-flight accounts of their experiences. Chapter 3 explains Experiment 1, including the set up of the VSL, and a variety of technical details about equipment and what we measured. Chapter 4 discusses the first experiment and recounts some of the problems we encountered on our mission to understand awe and wonder – problems in employing neurophenomenology, in designing the simulations and setting up the experiments. We discuss the lessons learned, and offer methodological improvements to support advancements in understanding consciousness and cognition. In Chapter 5, we review Experiment 2 and develop an explanation of the experiences of awe and wonder by integrating phenomenology, neuroscience, psychology, and information about cultural practices.

In Part II, we begin to explore a small part of the philosophical universe. Chapter 7 highlights what we learned from some of the limitations found in our approach, which focused on a positive category-based analysis. While this approach captures a lot of what we were after, it can miss important features of experience. Specifically it can easily overlook certain things that were left *unsaid* in the phenomenological interviews and how the participants' difficulty in expressing their experiences were in fact expressions of the unexpressible aspects of awe and wonder. In Chapter 8, we focus on how popular culture and media can shape the expectations of participants in regard to what they thought they would experience in the simulations. This is another way of highlighting the importance of taking cultural practices into consideration when doing cognitive science. Finally, in Chapter 8 we map out some of the philosophical implications of our approach to awe and wonder, and provide in more precise philosophical terms, an argument for a non-reductionist cognitive science.

Part I

Studying Awe and Wonder

2
Hermeneutical Explorations

This chapter looks at the original expressions and descriptions of the experiences of awe and wonder, as well as related experiences, including curiosity and feelings of humility, given by astronauts in their in-flight journals and in interviews after their return to Earth. Our aim in analyzing the astronauts' descriptions of their experiences was to identify and classify experiences that are related to, or that actually instantiate, awe and wonder. As mentioned in the previous chapter, we used two methods for textual analysis: hermeneutical analysis and computer-based syntactical analysis. We'll provide details of these methods and then discuss the results.

The texts actually constituted the starting point for our study. They consisted of 51 texts authored by 45 astronauts and cosmonauts either during their space travel (n= 17) – available at http://www.nasa.gov/centers/johnson/astronauts/journals_astronauts.html – or after their return to Earth (n= 34). These texts, totaling 23,000 words, captured details of their visual and affective experiences during space flight in the Space Shuttle or the International Space Station (ISS). The majority of the selected texts involved descriptions of experiences undergone while looking out of the space vehicle's windows with views of Earth or deep space from close earth orbit.

1 Syntactical analysis

For purposes of analysis, numbers from 1 to 51 were assigned to the texts and information about the author, including name, was excluded. In other words, we wanted the texts to speak for themselves without the potentially biasing context of demographics or personal identifiers. The first analysis we did was syntactical – looking at aspects of structure in regard to coherence and degrees of abstractness; the second was hermeneutical – focused on contextual meaning.

Our initial analysis made use of a computer-based tool – Coh-metrix (cohmetrix.com) –designed for syntactical analysis of texts and developed by psychologists at the Institute for Intelligent Systems at the University of Memphis. Coh-Metrix analyzes a number of metrics, including linguistic coherence based on a wide range of measures (Graesser et al. 2004). One of the things we wanted to know is whether there are significant differences between the in-flight journals and the astronauts' post-flight descriptions in interviews and books. Using this program, we looked at various patterns of syntactical structure in the language used by the astronauts to express their experiences. By this method, we found that the descriptions contained in the in-space journals were much more concrete (or less abstract) than the descriptions provided in later reflections and interviews.

Causal cohesion reflects the extent to which sentences relate to each other by expressed causal relations. It calculates the number of semantically identifiable causal verbs (e.g. *drop, fill*), causal particles (*because, in order to),* and semantically depleted verbs (e.g., *cause, make*). The more causal verbs in a text, the more the text is assumed to convey causal content. Cohesion drops, however, when a text contains too many causal verbs (signifying events and actions), but few causal particles that provide an indication of how the events and actions are connected. Causal cohesion is the ratio of causal particles to causal verbs (Figure 2.1a).

Coh-Metrix also measures word abstractness. A word is *abstract* when it has few distinctive features and few attributes. Abstractness is measured in Coh-Metrix by the noun *hypernym* values in Word-Net, an online lexical reference system (Fellbaum, 1998; Miller et al., 1990). The hypernym count is defined by Coh-Metrix as "the number of levels in a conceptual taxonomic hierarchy above (superordinate to) a word. For example, chair (as a seat) has 7 hypernym levels" (http://cohmetrix.memphis.edu/CohMetrixWeb2/HelpFile2.htm).

> seat → furniture → furnishings → instrumentality → artifact
> → object → entity

A word having more hypernym levels is considered to be more concrete. A word with fewer hypernym levels is considered to be more abstract (Figure 2.1b).

Likewise, Coh-Metrix assesses how concrete or non-abstract a word is on the basis of human ratings using the MRC Psycholinguistics Database

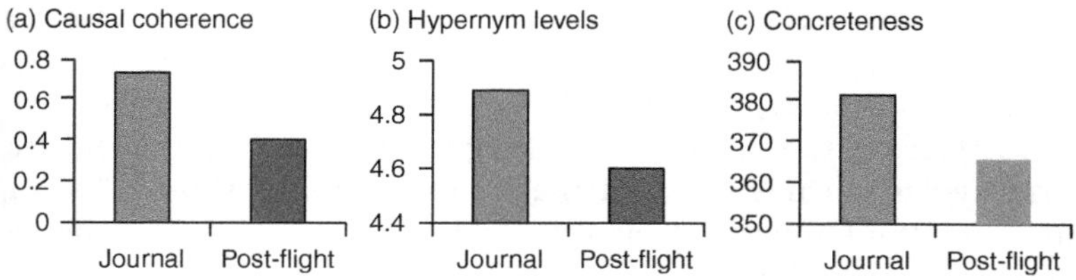

Figure 2.1 Coh-Metrix analysis of in-flight journal versus post-flight descriptions

(Coltheart 1981). Concreteness measures the degree to which a word has easily accessible mental images and direct sensory referents. Words like "spoon" or "water" are less abstract than words like "justice" or "moral". Higher scores are more concrete than low scores (Figure 2.1c).

On all three measures, the astronauts' in-flight journals reflect more causal coherence and are more concrete (less abstract) than post-flight reports. This signifies that they have the virtue of a more immediate, perception-based reporting. The journals were closer to the experience itself, not only in regard to proximity in time, but also in reflecting a certain experiential closeness to the perceptual event and the lived feeling that was involved in what the astronauts experienced. In addition, since memory and reflective thinking are involved in the post-flight texts, and include more temporal distance from the original experience, post-flight texts may not be as dependable as a veridical portrayal of that experience. This justifies a further analysis to see if these differences reflect differences in type or frequency of occurrence of the specific consensus categories (developed in the hermeneutical analysis, below), and whether the post-simulation interviews from our experiments are closer to in-flight journals or post-flight texts in this regard.

2 Hermeneutical analysis

In this section, we explain how we derived a set of categories to specify the precise experiences that the astronauts describe in their journals and reflections. As we'll show in subsequent chapters, these categories are used in the experiments in order to determine whether we were able to replicate the astronauts' experiences. The categories are put to use to interpret the follow up phenomenological interviews of subjects after they have experienced the simulated space environments. Both the interpretation of the astronauts' descriptions, and the interpretation of the participants' experiences during the experiments involve a hermeneutical approach. We offer a clarification of this approach here.

Hermeneutics has a long and complex history. It's usually defined as the theory and practice of interpretation. Traditionally, it was concerned with the proper interpretation of literary, sacred, and legal texts. In the twentieth century, however, the philosophical concept of hermeneutics broadened to include the idea that humans are, in Charles Taylor's phrase, "self-interpreting animals" (Taylor 1985). The use of hermeneutics as a method of textual interpretation, however, does not require that we enter into the larger philosophical questions about the possibility for human understanding more generally, found in thinkers like Taylor, Gadamer (1989), and Ricoeur. Rather, we can limit our concern to methodological issues (see Gallagher 1992 and 2004 for the broader philosophical discussion and for how hermeneutics relates to cognitive science, respectively).

2.1 Some methodological notes

For our purposes, we can consider a few important hermeneutical issues involved in the interpretation of any text. First, we note that any interpreter carries with her certain biases and interests that need to be acknowledged. Consider, for example, a juror who is predisposed to think (or is already convinced) that the defendant is guilty. She will most likely interpret any evidence presented in the case as evidence for guilt. She would have to work harder to maintain an objective perspective on the facts presented. In a similar way, if we read a text in order to find expressions of the feeling of awe, we likely already have a concept of awe in mind that predisposes us to consider certain expressions as expressions of awe. We can call this the *definitional bias*. Practically speaking, it's impossible to read and understand a text without already having some presuppositions that shape our expectations of what we'll find in the text. If we don't know what we're looking for, we'll have a very difficult time saying whether we've found it or not. In order to mitigate this bias, we need to identify how we define the things that we are looking for. This is the reason we needed to specify our working definitions of awe and wonder, although they remained fairly abstract.

- **Awe:** a direct and initial experience or feeling when faced with something amazing, incomprehensible, or sublime
- **Wonder:** a reflective experience motivated when one is unable to put things into a familiar conceptual framework – leading to open questions rather than conclusions

Awe motivates wonder, and wonder has the potential to change one's perspective on life. These definitions turned out to be consistent with some others we found subsequently. Fuller (2009), for example, argued that wonder bridges emotion with the desire to apply order to the universe. This repeats the sentiment of Albert Magnus (1988, 557), who a millennium ago stated, "...wonder is the movement of the man who does not know on his way to finding out." Both awe and wonder have been classified as emotions, ambiguous states associated with surprise and fear, as well as peak experiences that can be life transforming (Shiota, Keltner and Mossman, 2007; Ekman 1992; Emmons 2005). Keltner and Haidt (2003) suggest that awe requires perceived vastness and a need for accommodation. Perceived vastness is described as being an experience that is considered both powerful and moving (Keltner and Haidt 2003). This could relate to any perceived objects as being larger than oneself, involving physical size, social status, or other categories involving magnitude (Bonner and Friedman 2011). Furthermore, a need for accommodation is described as related to the inability to incorporate an experience into current mental structures (Keltner and Haidt 2003). As a result, accommodation requires mental reorganization in order to understand the experience (Bonner and Friedman 2011). Bulkeley (2002) describes wonder as being an abrupt decentering of the self when faced with a novel and powerful experience and a recentering of the self in response to new knowledge and understanding. Both awe and wonder may require or motivate further cognitive processing to comprehend what is experienced, which suggests that they may involve cognitive aspects as well as emotional ones (See Chapter 6 for further discussion of how awe and wonder are defined in the tradition).

In addition to awe and wonder, we looked at the closely related concepts of curiosity and humility. Curiosity also involves a desire to piece things together, but in a different way to wonder. Curiosity involves wanting to know, see, experience, and/or understand *more*. The object of this wanting may be technical, logical, moral, or existential. John Milton McIndoo (2014) opposed curiosity to the impulse to flee in fear. That which may incite fear at first, may become intriguing, as familiarity grows. In this respect, curiosity, which is "world-oriented," acts as an important contrast to humility, which is "self-oriented." Theoretical views vary greatly concerning the nature of humility; it gets attached to everything from psychological concerns of self-esteem, cultural roles, and the limits of knowledge about the universe. Regardless, humility demands a sense of perspective, where one must place oneself in scale to someone or something else. In the present study, humility is a sense one has about one's relation to the universe (an issue of scale) or one's significance (an issue of moral aspect).

Having definitions, however, leads to a second issue. Even when we have a definition of what we are looking for, we have to decide whether what we find actually meets the definition. For example, we already need to know what we mean by the terms "amazing" or "sublime", which occur in the definition of awe, and know whether what we see described is a description of something amazing or sublime.

These issues lead too easily to the idea that interpretation is simply subjective. There are a number of things to note and a number of strategies available to prevent interpretations from being merely subjective. First, if this were the way interpretation worked, then scientific interpretation would be equally subjective. In science, we like to talk about data (and we will talk about data); but one always needs to interpret data since it is not self-explanatory. In the interpretation of scientific data, one might already be convinced by a certain theory or hypothesis, and like the juror, one might simply read the evidence in a way that makes it fit within the framework of that theory. There are, however, safeguards in science that prevent this kind of interpretation from reducing to purely subjective judgment. There are controls, and there are methods that allow for replication. Importantly, science appeals to intersubjective verification. That is, scientists communicate their results to other scientists to be further tested. For science, in fact, these are hermeneutical safeguards. We can appeal to similar methodological safeguards.

With respect to textual interpretation, we cannot make the text in question say anything we want. The text pushes back, so to speak. I can't pick up a dialogue by Plato and somehow interpret it to be a play by Shakespeare. To get more precise interpretations, I can do some historical research to understand the context that (literally) goes with the text. Textual interpretation involves what is usually called the "hermeneutical circle." This principle can be expressed as follows: one part of the text throws light on another part; the whole of the text constrains the interpretation of the parts, and vice versa – the interpretation of parts constrains the interpretation of the whole. The idea is that the interpreter needs to go back and forth between parts and whole until the understood meaning of the parts coherently mesh with the understood meaning of the whole. Again, however, this is not a subjective process. The text itself challenges the interpreter and in this process can reveal the reader's biases.

In terms of the astronauts' journals and post-flight reports, we can consider the various contexts within which they were written. The fact that some of the texts are in-flight journals, that some are post-flight reflections, and that some are responses to certain questions during an interview are aspects that may help to understand what we are reading. Indeed, given what the syntactical analysis told us about the contrast between in-flight journals and post-flight reports, we can weigh some of these texts as closer to the experience in question. We can also adopt the scientific safeguard of using intersubjective verification. We can have multiple interpreters looking at the texts in order to have intersubjective confirmation of a consensus interpretation.

2.2 An initial sorting

Initially, one person read through a large set of texts published as in-flight journals on the NASA website and other post-flight accounts by astronauts to determine whether they included material that generally fit the description of reports on "*experience of interest.*" In most cases, this was quite easy to discern. For example, a number of in-flight journal entries are simply reports on events that happened during space flight or reports on technical details about equipment or experiments. For example, Flight Engineer, Sandra Magnus, reports in a journal entry entitled, "Food and Cooking in Space, Part 2:"

> It turns out that with duct tape (yep, the ubiquitous duct tape) and plastic bags you can do just about anything. I have definitely been on a learning curve. With each successive experiment I am getting less and less messy. (Note I am not saying I am getting more and more clean and orderly, just less messy!) From day one, when I realized how useful and necessary plastic bags are to the cooking process I have been on the look out for the right size (and cleanliness) of plastic bags. (http://www.nasa.gov/mission_pages/station /expeditions/expedition18/journal_sandra_magnus_7.html)

One might think that she is reporting on experience. She's telling us about her experience with duct tape and with cooking in space. But this is not what the phenomenologist considers experience in the strict sense – that is, the kind of experience we are interested in. Here we can appeal to a bit of philosophical German – the distinction between *Erfahrung* and *Erlebnis*. Both words can be translated as "experience." If I were to say that I have lots of experience with cooking, I would use a variation of the word *Erfahrung*. It would signify the type of experience one might talk about if one were answering a job advertisement for a chef. I've cooked for ten years, therefore I have lots of experience. We get closer to the notion of experience that we are interested in with the term, *Erlebnis*, which is often translated as "lived experience." Consider what someone might say about cooking: e.g., "Cooking in space is a real challenge, but *I love doing it*; it brings a *sense of accomplishment* and *I feel fulfilled* by my ability to create a dish out of unexciting ingredients." In this case, the person would be describing what it's like to engage in this kind of project – what he feels like when he is cooking, and what it does for his sense of accomplishment. In this case, there are feelings being expressed, and descriptions of how the person experiences a particular activity. So the difference between experience as *Erfarhung* and lived experience as *Erlebnis* is like the difference between, "I am doing X," and "I really enjoy (or hate, or am bored with) what I am doing."

Let's further note, however, that we were not interested in all lived experiences that the astronauts had. For example, we were not interested in whether or not they really enjoyed cooking. Rather, we were looking for experiences that came close to what we were defining as awe and wonder, and we didn't find any experiences of awe and wonder connected with kitchen duties, even in outer space. Consider, then, what the same Flight Engineer, Sandra Magnus, reports in a journal entry entitled, "The night pass." She describes looking out of the ISS windows at night.

> The night sky, the heavens, though is what really catches the eye. Even though the Earth's horizon is dark, light provided by the clouds and the city lights reflecting off of the clouds, provides enough illumination to discern the difference between the Earth and space.... You are swimming in a sea of beautiful lights that can only be seen in the dark. As you gaze at the multitude of points glittering in the night, it is hard to imagine that each one is a world or worlds or stars like our sun. They are so remote and seem so tiny. The vastness of space is truly evident as you watch the Earth turn slowly beneath. It is awe inspiring and overwhelming all at once and oh, so beautiful! (http://www.nasa.gov/mission_pages/station/expeditions/expedition18/journal_sandra_magnus_10.html).

There are various words and phrases that speak of her lived experience, like "catches the eye," "swimming in a sea of beautiful lights," "awe inspiring," and "overwhelming," and tells us something about an experience that could easily be associated with awe – indeed, she uses the term "awe." Even if she didn't use the word "awe," we should still be able to recognize certain aspects of what she is describing as related to our definition of "awe." The idea that something is "overwhelming" (in a good way[1]) expresses an experience that seems close to awe. It is just at this point, however, that one may ask why a term like "overwhelming" is connected with awe. Two things can be said in this regard. First, Sandra Magnus herself says it. That is, she associates the terms "awe inspiring" and "overwhelming." The claim here is not that Sandra Magnus is an expert on the meaning of awe; rather the important thing here is that we are paying attention to what the astronaut, as an experiencer, is saying about her experience. Second, we can find some intersubjective verification that the feeling of being overwhelmed is closely associated with the experience of awe, both in the reports of other astronauts who make the same associations, and in the analysis of numerous interpreters who are given our working definition of awe and who pick out feelings of being overwhelmed as lived experiences that instantiate that definition.

It's important to note that not everyone who describes the lived experience of being overwhelmed will say that they "were overwhelmed." There are different ways to say the same thing, or at least, to express an experience that is close enough to the feeling of being overwhelmed that a number of people would judge it to mean something very similar. So someone might say that the view of the night sky out the window was really "too much," or "more than I could take," or that "my senses were overloaded." We would count these expressions as statements of being overwhelmed.

One final note. One could easily think that there might be some relation between feeling overwhelmed and expressions of feeling small in contrast to something extremely large. That seems right. Sandra Magnus says, "As you gaze at the multitude of points glittering in the night, it is hard to imagine that each one is a world or worlds or stars like our sun. They are so remote and seem so tiny. The vastness of space is truly evident as you watch the Earth turn slowly beneath." We'll see that this is a theme repeated by a number of astronauts. We refer to the specific changes in the feeling of relative size as "scale effects" – for example,

---

1 We note that awe is not always or necessarily a good experience. Awe can be 'awe-ful', as in *awful*. A negative form of awe (something that might come with witnessing an explosion, for example) might be paralyzing. If awe is not necessarily positive, it usually is, and all of the experiences expressed by the astronauts, as far as we can tell, were positive.

the sense that the universe is vast and that you are so small, or the earth is so small. In some cases, one might feel overwhelmed by the vastness of everything – an immediate feeling brought on by what you actually see. In other cases, you may feel overwhelmed just by thinking that the universe is so huge. The combination of feeling overwhelmed and experiencing scale effects is sometimes equated to the experience of the sublime. In any case, although there seems to be a close connection between feeling overwhelmed and scale effects, there are ways of feeling overwhelmed without experiencing scale effects – e.g., when you experience a sensory overload, or in an aesthetic experience of beauty that does not involve scale effects. Accordingly, in the categories that we develop, we distinguish the feeling of being *overwhelmed* from the experience of *scale effects* and the experience of the *sublime*. The close relation between these experiences, however, also suggests that they are close variations of the experience of awe. To be able to say, however, that in one case I experience awe by feeling overwhelmed, or by experiencing scale effects, adds important specification to the general conception of awe.

2.3 Consensus categories

Once we identified a set of texts to work with, the real work of interpretation began. Two "primary" interpreters, working independently, read the complete set of texts and conducted an interpretive analysis that attempted to identify categories of experience that fit the definition of awe and wonder. We were also interested in expressions of curiosity (which relates to wonder) and humility (which relates to awe). After their independent analyzes, the primary interpreters combined their results and identified a total of 48 categories of experience expressed in the texts that appeared to be related to awe, wonder, curiosity or humility (AWCH). There was already a good agreement on the list of categories. But to verify the categories, we gave different subsets of the complete set to 20 other, "secondary" readers (15 graduate students in philosophy and cognitive science, and 5 undergraduate students) along with our 48 categories, and asked them to identify instances of the categories in the subset of texts they were reading. The idea was to gain some inter-rater reliability. In this manner, we were able to verify a "consensus" set of 34 of the 48 original categories (see Table 2.1).

The consensus categories helped to make more precise the general experiences of awe, wonder, curiosity, and humility. We were able to specify the following operational definitions in terms of the 34 categories found in the texts.

Table 2.1 The 34 consensus categories of experiences had by astronauts

- Aesthetic appreciation
- Captured by view/ drawn to phenomenon
- Change (internal or bodily change)
- Connectedness (feeling connected with something without losing distinctness)
- Contentment (tranquility, feeling relaxed or at peace)
- Disorientation
- Dream-like experience (feeling of unreality, abstract feeling)
- Elation
- Emotional (general emotional feeling or arousal)
- Experience-hungry (wanting more of a particular experience)
- Exteroceptive intensive experiences (sensory overload, silence)
- Floating (bodily, feelings of weightlessness)
- Floating in void (not related to weightlessness)
- Fulfillment
- Home (feeling of being at home)
- Inspired
- Intellectual appreciation (for order, analysis, complexity)
- Interest/inquisitiveness
- Interoceptive intensive experiences
- Joy (feeling of happiness)
- Nostalgia
- Overwhelmed
- Perspectival (spatial) change
- Perspectival shift (internal change of [moral] attitude)
- Peace (conceptual thoughts about)
- Pleasure
- Poetic expression
- Responsibility (towards others)
- Scale effects (feelings of the vastness of the universe or one's own smallness/ insignificance)
- Sublime
- Surprise
- Totality (wholeness of what is experienced; big picture)
- Unity of external (earth, universe, people on earth, interrelatedness)
- Unity with whole (feeling of oneness with; holistic feeling)

Awe: A direct and initial feeling when faced with something incomprehensible or sublime.

Specification: Captured by view/ drawn to phenomenon; elation; experience-hungry, overwhelmed, scale effects, sublime, surprise.

Wonder: A reflective feeling one has when unable to put things back into a familiar conceptual framework

Specification: Inspired; Perspectival shift; Nostalgia; Unity with whole; Unity of external; Responsibility.

Curiosity: Wanting to know, see, experience, and/or understand more.
Specification: Interest/inquisitiveness; Experience-hungry; Intellectual appreciation.

Humility: A sense one has about one's relation to one's surroundings or of one's significance.
Specification: Responsibility; Unity with whole; scale effects.

The specification that we achieved following this method is important. We started with working definitions – a hermeneutical guide to what we were looking for. With the consensus categories, however, we have something different. Not only do they add specification, defining our understanding of AWCH even more; importantly, they are derived from the astronauts' own descriptions of experiences they had as they looked out of the Shuttle or ISS windows at either the earth or deep space.

There are further differentiations to make across these consensus categories. We can cut the cake a number of different ways and through a number of different layers. For example, some of the experiences are clearly *aesthetic* – obviously those we categorized as aesthetic experiences, but also experiences of the sublime, and poetic expression. In addition, sometimes the aesthetic experience leads to inspiration; sometimes it's beauty that motivates the experience of being overwhelmed. Other experiences may be purely *intellectual*. Some involve a kind of *spiritual* change where spiritual indicates some aspect of self-transformation inspired by these experiences. Some experiences are expressed in *religious* terms. For example, Jeff Williams, who acted as an advisor on our project, in his own in-flight journal wrote: "You can never tire of looking at the part of God's creation we call Earth." Looking at the earth from space can also motivate experiences that involve environmental (ecological) and/or *ethical* concerns. These various attitudes cut across the categories of awe and wonder. In certain contexts, it will be useful to refer to these attitudes; but there is no suggestion that any of them exclusively correlate with the basic categories of AWCH, or with the consensus categories.

3 The texts themselves

In this section, we'll provide some sample texts and examples of the hermeneutical analysis. The complete texts from the astronauts' journals can be found at the NASA website (http://www.nasa-usa.de/centers/johnson/astronauts/journals_astronauts.html).

The first example is taken from the astronaut Greg Chamitoff's in-flight journal.

> Incredible! I just noticed we were approaching London around midnight GMT. I decided to turn off all the lights and set myself up for some hopeful night shots. What an amazing, spectacular, incredible, mind blowing view! So for a moment I just stared at the incredible display of life below me. From there we flew across the rest of Europe in a few minutes and I was just overwhelmed with the beauty of our civilization as it was, splattered across the dark landscape.

The idea that Chamitoff "sets himself up" to have these experiences might suggest that there is some fit for the category of being *"experience hungry"* – which is the idea that the person's interest in having these sorts of experiences is so strong that they proactively put themselves in a position to have them. This is not clear, however, since he is setting up to take photographs out the window, and it seems that it's more a case of him being *surprised* by what he sees. Clearly, however, there is an experience of being *overwhelmed*: "What an amazing, spectacular, incredible, mind blowing view!" The next sentence might be a case of being *captured by the view/drawn to the phenomenon,* although it may depend on how momentary his staring at it was. For that reason, we did not categorize it in that way. The last sentence, however, clearly indicates being *overwhelmed* again and an experience of *aesthetic appreciation*.

Here's another passage from Chamitoff's journal.

> I went up to the flight deck to see the view, and wow, it was incredible [surprise; overwhelming]. The first sensation of looking out the window was very disorienting. Everything seemed to be floating – me, the shuttle, and the Earth, and all in different orientations.

Again, there is an element of *surprise* and being *overwhelmed* signaled in the words "incredible" and "wow." He explicitly mentions *disorientation* and *floating*. This includes his own weightless floating as well as a more general floating of everything – a *floating in the void*. We will come back to the issue of weightlessness and to what extent it might be a causal factor in some of these experiences. This turns out to be a limitation in the experiments we conducted, but in the end we determined that there was no direct causal link between weightlessness and the experiences of AWCH. Indeed, our hypothesis was that in the cases described by the astronauts, the experiences are more directly caused by the views out the windows of the ISS or Space Shuttle. We can look at one more example from Chamitoff's journal to make that clear.

Chamitoff describes an experience that happens as he gazes out of the JEM (Japanese Experiment Module) windows.

> At the far end of the JEM, there are two very large, port-facing windows, which are awesome. Most other windows on the station point down, which is great for Earth observations and photography, but you can't get a big picture perspective from that view. The JEM windows face the horizon, and the views are incredible. No doubt I'll be spending much of my free time gazing out of those windows, looking over the Earth, and just wondering what it all means.

In this case, we could interpret that last sentence as indicating that he is *experience hungry* for the amazing feelings that he has already experienced, and that part of what he gets out of it is this experience of wonder – in the sense of an intellectual wondering about what it all means. He later explains his motivation and interest in the window.

> Something else happened recently that will always be a special memory for me. I took a peek out the side-facing JEM windows one evening, without camera in hand, and was so mesmerized that I ended up gazing upon the Earth for an entire 90-minute orbit. Believe it or not, that is the first time I have done that. A hundred times I thought, "I should go grab the camera," but I decided to just try to capture this one orbit with my own eyes and burn it into my brain....

This is a clear instance of being *captured by the view* or *drawn to the phenomenon*. We find a similar experience expressed by Scott Carpenter.

> I found it difficult to tear my eyes away and go on to something else. Everything is so new and so awe-inspiring that it is difficult to concentrate for very long on any one thing. Later on, when I knew that I was returning to some wonderful sight that I had seen before, I could hardly wait to get there. Using the special camera I carried, I took pictures as fast as I could, and, as I raced towards night...I saw the beginnings of the most fantastically beautiful view I have ever had – my first sunset in space....At every new sight, my elation was renewed, and I kept waiting again for the next one.... (Carpenter 1962, p. 450)

Another astronaut, Peggy Whitson, in her in-flight journal, in the form of a "letter home," expresses a different experience.

> There is no way that I can imagine, especially after seeing our planet from this vantage point, that bringing our cultures closer together and proliferating understanding in our differences as well as our similarities, can be a bad endeavor.

Whitson indicates what we call a *perspectival shift* – a dawning of a moral sense that is tied to a visual experience of seeing earth from space; a sense that can lead to a renewed responsibility towards others or understanding of other cultures. This kind of perspectival shift can motivate a personal rethinking that extends to ethical and political considerations. Another good example of this kind of reflection can be found in Russell Schweickart's description of viewing earth while on the Apollo 9 mission.

> You identify with Houston and then you identify with Los Angeles and Phoenix and New Orleans. And the next thing you recognize in yourself is that you're identifying with North Africa – you look forward to that, you anticipate it, and there it is. And that whole process of what it is that you identify with begins to shift. When you go around the Earth in an hour and a half, you begin to recognize that your identity is with that whole thing. That makes a change. You look down and you can't imagine how many borders and boundaries you cross, again and again and again, and you don't even see them. There you are – hundreds of people in the Mideast killing each other over some imaginary line that you're not even aware of and that you can't see. From where you see it, the thing is a whole, and it's so beautiful. You wish you could take one in each hand, one from each side in the various conflicts, and say, "Look, Look at it from this perspective. Look at that. What's important?" (Cited in White 1987, pp. 11–13)

So far, we have been placing reports of experiences into the various categories that we have taken from the astronauts own words. In some sense, one might think, this should be no problem since indeed, the experiences themselves are the basis for the consensus categories. We think that we've given a sufficient number of examples, however, to show that this is not just a kind of mechanical application of categories. Some of the reports on experience are nuanced in a way that leaves it uncertain whether they fit within a particular category or not. Beyond that, however, there is another issue that will take on some importance later in our analysis. Peggy Whitson gives some indication of this issue. This is the idea that it is sometimes difficult to find words to express precisely what these experiences are. In some cases, the lack of words is "expressed" precisely by a lack of words. Here, however, Whitson explicitly puts the lack of words into words.

> Being here, living here, is something that I will probably spend the rest of my life striving to find just the right words to try and encompass and convey just a fraction of what makes our endeavors in space so special and essential.

This statement expresses a generic kind of reflective wonder, but is not easily placed within any of the consensus categories. One might add a category – call it *ineffability* – pointing to these kinds of explicit statements about it; but the difficult task would be to use this category to indicate a situation in which something remains unsaid, because there are no words to express the experience.

3
Awe and Wonder in a Simulated Space Flight: Experiment 1

This chapter explains the experimental design and the results of our first experiment, the design of the simulated environments used in that experiment, and a variety of technical details about the equipment we used in addition to what we measured. The use of a simulated environment to create an experimental test bed where a number of variables can be easily manipulated is an appealing idea for the exploration of certain hard-to-test behavioral and experiential phenomena. For several reasons, the study of such experiences without the use of simulation would be unfeasible or too expensive. For example, testing subjects on Earth, rather than sending them into space, is more affordable and also provides a degree of control over a number of important aspects. As will be illustrated, however, the use of simulation (instead of real space flight) also comes with certain limitations.

1 Designing space on earth

The challenge was to create an environment within a lab setting that would allow us to recreate the experiences of space travel. The engineers and artists on our team at the Institute for Simulation and Training used the latest simulation technology to design and construct two environments that could act as test beds for our experiments. In this chapter, we focus on the first experiment, and we'll describe the second experiment in Chapter 5. The simulation for the first experiment was a mixed-reality test environment – the "Virtual Space Lab" (VSL) modelled on an International Space Station (IIS) workstation. A mixed-reality simulation combines virtual and real elements. The VSL was designed to expose subjects to simulated stimuli of the earth and deep space while sitting in an environment similar to those experienced by astronauts. We'll focus here not only on the technical details, but also on the content of the simulation – what precisely were our participants getting into when they entered the VSL?

We decided that the basis for the virtual part of the simulation had to be realistic, dynamic images of space representing what astronauts would experience during their real space flights. Our first step was to identify the images to be used as the basis for the simulation. The choice of images was informed by a number of considerations, including availability, image quality, details about practices in the different space vehicles involved, and historical analysis of the image. In regard to the latter, for example, the classic "Blue Marble" image, the best-known example of a space image that strongly affected the feelings of many people worldwide, is a photograph taken on December 7, 1972 by the Apollo 17 mission. This image shows the earth from a perspective in outer space and was the first photograph many people saw of the planet they inhabited. Its effect is dependent upon the powerful emotions it evokes. Before it was able to have this kind of impact, the photograph – taken with a 70 mm Hasselblad camera and an 80 mm lens – had to be reoriented. Originally, the South Pole was visible to the crew of the Apollo 17 at the top of the globe; the published image was rotated 180° to provide a view familiar from maps and atlases. In effect, what the astronauts actually saw and marvelled about was not precisely depicted in the popular image. In addition, as a rule, the earth appears partially in shadow, so the most celebrated image of the planet is also a relatively uncharacteristic one. This example highlights the fact that the task of finding images suitable for a simulation cannot rely exclusively on the reports from the astronauts or an uncritical acceptance of images from NASA.

As we indicated in Chapter 1, our research team included members of the *Bildakt* group, art historians, and philosophers working at the Humboldt University in Berlin. They used the NASA Image Database to define criteria for image selection for the VSL (see http://www.nasa.gov/multimedia/imagegallery/index.html). The NASA database is classified according to four major categories (item, location, person, time). These categories do not differentiate between media, contexts, or individuals, so that specific keywords like "earth" (23,491 entries) or "sun" (16,092 entries) also return a large number of diagrams, models, artistic views (drawings, paintings), documentary photographs of laboratories, staff, etc. It turned out that less than 10% of all images in the database were relevant for our project. The following selection criteria were defined:

- no (or little) artifacts due to optical refraction and other lens effects;
- no aged colors on chemical films;
- no visible manipulation in terms of coloring, digital editing, etc.;
- no markings and inscriptions (except watermarks made by a photographer, which can be cropped);
- pictures should be focused or rich in detail;
- views should be possible from a spacecraft within the earth–moon system and could be made by human observers;
- earth view: Earth should be visible as crescent, not as iconic blue marble;
- space view: no particular object(s) should dominate the image; image may be used as background for animation; number of visible stars should be higher than in a sky seen from earth (clarity of sight).

A number of challenges presented themselves in the attempt to follow these criteria. First, finding high-quality images that showed on monitors in high fidelity was difficult. Our first attempts looked flat. Second, the images as used in the simulations had to be dynamic in ways that did not violate the laws of physics.

One part of the first problem was solved when, fortunately, on 25 January 2012, NASA released a composite image of the western hemisphere of the earth entitled Blue Marble 2012 (Figure 3.1 A). The picture was widely viewed, logging over 3.1 million views on the Flickr image-hosting website within the first week of release (see https://www.flickr.com/ photos/gsfc/6760135001/). On February 2, 2012, NASA released a companion to this new Blue Marble, showing a composite image of the eastern hemisphere. These images served as the basis for developing a dynamical simulation of the rotating earth for the VSL – an image that moved relative to the observer. Simulation developers also introduced a crescent element (Figure 3.1 B, modified image).

A number of important limitations constrained the design and construction of the VSL. The VSL was constructed in a lab in Florida. It was obviously located on the earth and not moving relative to anything on the earth. One clear limitation was that we could not simulate microgravity in any way that would not interfere with the visuals and the experimental requirements. However, this was not expected to be a problem for several reasons. First, reports in the astronauts' journals suggest that there is little explicit connection between weightlessness and the visual stimuli that

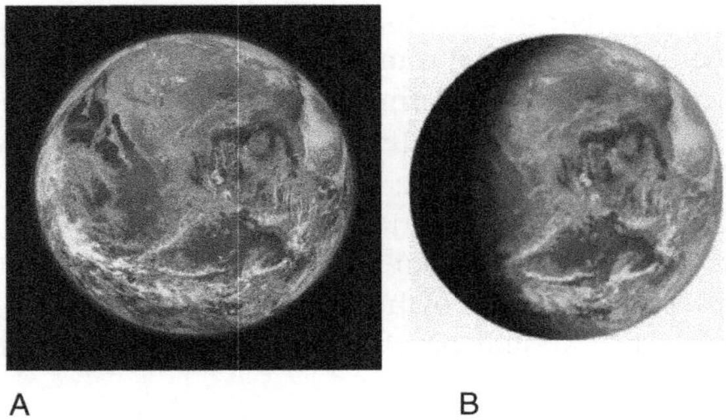

Figure 3.1 (A) Blue Marble 2012. The image is taken from the VIIRS instrument aboard NASA's Earth-observing satellite – Suomi NPP. It's a composite image that uses a number of swaths of the Earth's surface taken on January 4, 2012. (B) Modified image used as the basis for one part of the simulation.

generated the experiences in question. Weightlessness is an issue at the beginning of the space journey and is addressed in a pragmatic fashion with the main concern about movement and being able to control action. After a few days, these issues are resolved and there is not much discussion of weightlessness and, with one exception, any mention of weightlessness in connection with the visual experiences at the window. Second, we employed a seated workstation scenario that minimized the different practical effects of microgravity *versus* gravity. That is, in the circumstances of being strapped into a workstation, as astronauts sometimes are, the presence or absence of gravity makes little practical difference.

It is important to note, nevertheless, that we were not able to control for the implicit effects of microgravity and vestibular modulations on the visual system (Clément 1998; Clément, Lathan and Lockerd 2008; Harris et al. 2010). There is some evidence that long-term weightlessness does have physiological effects on vision (Kramer et al. 2012; Mader et al. 2011). However, such changes manifest themselves only after several months in microgravity. Of the 38 astronauts who have lived aboard the ISS for 3–6 months, 7 of them have developed measurable ophthalmic changes after flight, including optic disc edema, globe flattening, choroidal folds, hyperopic shifts, and cotton wool spots (Zwart et al. 2012). Since these are long-term changes that take effect after returning to Earth, it is unlikely that they would affect visual experiences during space flight.

A more significant worry would be for any short-term effects of microgravity on vision (Clément 1998; Clément, Lathan and Lockerd 2008; Villard et al. 2005). Cheron et al. (2014), for example, report a difference in top-down modulation of vision-related processes for some stimuli in weightless conditions (on the ISS) compared to in Earth's gravitational field. This applies only to a visual stimulus that involved a navigational task (a 3-D virtual tunnel that subjects were asked to navigate), and did not apply to the presentation of an image that did not call for movement or a navigation task on the part of the subject (a 2-D classic checkerboard stimulus). As the authors explain, the difference is likely generated along the dorsal pathway (the visual pathway specific for action preparation and motor control) and therefore tied to the motor task. That's understandable given that movement in microgravity is different than in gravity. In this respect, since our experiments did not require motor tasks, we would expect no such modulation difference in the microgravity condition.

A number of studies of the visual system during space flight suggest few significant changes in visual function: "visual performance was neither significantly degraded nor improved during space flight.... [V]isual performance testing on Mercury, Gemini and Apollo missions revealed few significant changes in visual function.... Except for contrast sensitivity, in-flight experiments on a group of astronauts revealed no statistically significant changes in any of the parameters measured compared to pre-flight baseline" (Clément and Reschke 2008, p. 104; see Task and Genco 1987). Contrast sensitivity declined to 60% after 5 days. "Even at these levels of change, it was concluded that the effect of space flight on visual function was relatively small" (Ibid., see Nicogossian and Parker 1982).

We note that, not only the effects of microgravity, but also the complex physical demands of space flight, in-space work environments and intensive work schedules can involve significant and complex bodily changes that may affect cognition. Changes in circadian rhythms, hormonal changes, and disrupted sleep schedules can lead to decreased cognitive performance (Mallis and DeRoshia 2005; Newberg 1994; Strollo 1999). Variations in such conditions, however, are not regular and depend on circumstances and the physical conditions of each individual astronaut (and likewise each individual participant in the experiment). In some cases, a participant reported that he or she had not had a full-night's sleep prior to the day of the experiment; in some cases the individual reported stress situations related to work or schooling. In some of these cases, the participants reported experiences of awe and wonder during the simulations and in some cases, they reported no such experiences.

Another set of limitations involved what we might term stage setting, and it contributed to considerations about VSL design. Obviously, for the experiment, we couldn't simply push participants into the VSL and see what happened. We had to talk them into it, so to speak. That is, we had to situate them in a narrative that would be convincing, without letting them know what we were looking for. In that sense, there was a narrative (rather than just a physical) entry into and exit from the simulation. Participants, of course, knew that they were not leaving Earth and that they would not really be in outer space. Yet, we thought, a higher level of immersion in the simulation would depend on making the experience as convincing as possible, given certain physical and budgetary limitations. Accordingly, we designed launch and landing sequences to facilitate immersion in the narrative of being in space. We introduced convincing auditory effects to simulate launch – the loud and authentic sound of the space shuttle launch rockets that began after a vocal countdown and eventually cut-off to silence, which signalled arrival in space. This was a simple solution that worked quite well. A number of participants reported a high degree of realism connected with the sound and vibrations and a feeling that they were taking-off. The landing sequence, which we considered less important, involved a series of communicative radio announcements, much in the way that Houston Mission Control communicates with astronauts in space.

The physical size and location of the VSL also required some problem solving involving lab space and noise levels. The interior of the VSL was modelled on a workstation on the ISS which allowed access to a console of computer equipment and windows (see Figure 1.1 in Chapter 1). In the VSL, we wanted to limit any distraction from the window visuals, so we minimized the complexity of the console area and provided only a desktop-computer monitor, (which was kept completely dark during the time the windows were open), and a desktop mouse and keyboard. (Figures 3.2, 3.3, and 3.4).

Two large-scale monitors were located so that they were framed by VSL windows. They were installed in such a way that they excluded any exterior light from reflecting on the monitors. Virtual portholes could be opened or closed to allow access to space views in the windows. Four simulations were designed.

1. The earth (based on the modified image of Blue Marble 2012, Figure 3.1B, above).
2. Deep space (composed of distant stars and gassy formations).
3. The earth-with-object (the earth as in (1) plus the passing appearance of the ISS).
4. Deep space with object (deep space as in (2) plus the passing appearance of the moon).

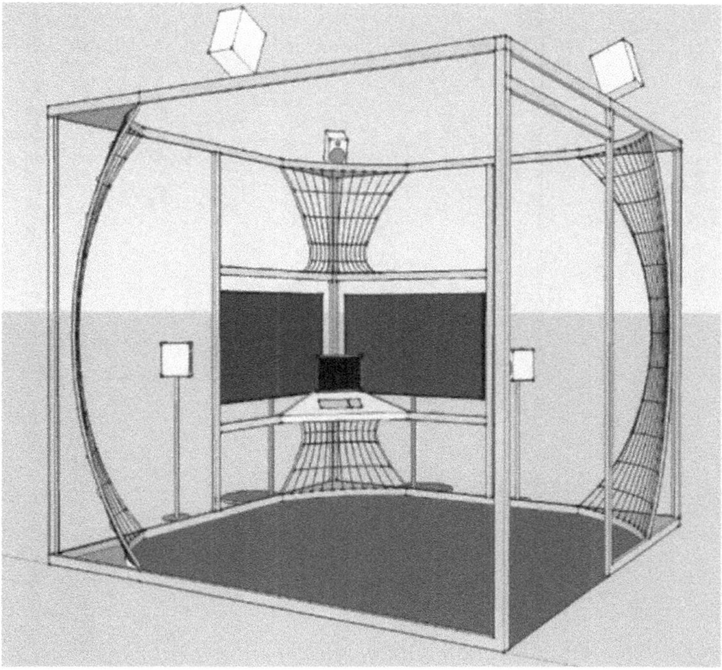

Figure 3.2 Frame view of the VSL. White boxes are audio speakers.

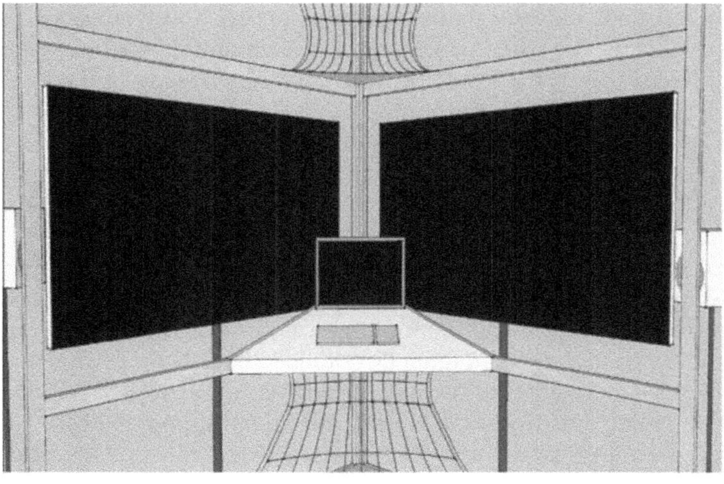

Figure 3.3 Close-up frame view of VSL showing two large-scale monitors to be located behind windows, and workstation.

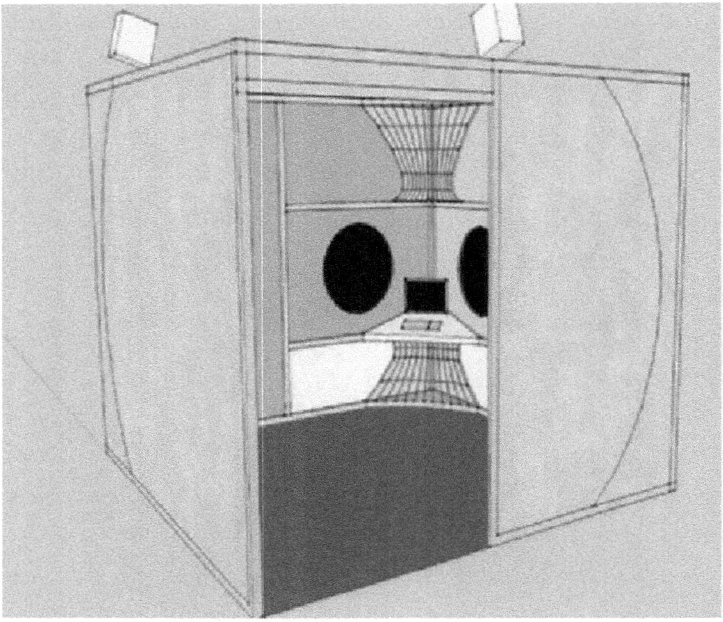

Figure 3.4 VSL with walls and windows added.

Objects were introduced into conditions (3) and (4) because, although astronaut texts indicate that experiences of awe and wonder occurred when viewing the earth or deep space, astronauts' missions occurred in different spacecrafts, some of which allowed viewing of the ISS, and some of which were views from the ISS. We thought it would be important to include these variations.

2 The first experiment

Psychological methods for studying experiential states rely on self-report questionnaire data that are practical, readily quantifiable, and easy to interpret. This method allows for large amounts of data to be collected and analyzed in a timely manner (McDonald 2008). Despite these advantages, self-report questionnaires are subjective and can be influenced by the participants' willingness to participate, response biases, acquiescent responding, as well as extreme responding that can have a significant effect on the relaibility and validilty of these measures (Paulhus and Vazire 2009). A survey, however, inevitably can only approximate experience and the adequate testing for experiences of awe and wonder can be especially difficult. Additionally, self-report state based questionnaires are given after a stimulus is presented. Consequently, memory errors that cause skewed responses may occur due to factors that intervene between the stimulus presentation and the adminisitration of the questionnaire (Tourangeau 2000). Overall, self-report questionnaires are practical and are consistently used by researchers, but are limited due to the subjectivity of participant responses.

Neuroscience methods for studying experience rely on psychophysiological systems that record metabolic and electrical signals from the body (Wubbels et al. 2007; Gevins et al. 1975). The majority of these systems have advantages of being objective, unobtrusive, and continuous, which allow for signals to be processed in real-time. However, psycho-physiological systems are subject to extraneous noise, temporal, and spatial resolution limitations, and are difficult to interpret due to the large number of dependent variables outputted per system (Dirican and Gokturk 2011). Moreover, at best, they are indirect measurements of experience and difficult to interpret in that sense as well. In general, psycho-physiological systems are beneficial in that they are unbiased and objective, but the information that is collected is challenging to understand without other forms of data such as performance results.

Phenomenological methods for studying experience in others rely on interviews that are used to evoke a description of an experience from the unique perspective of the individual (Dukes 1984). Phenomenological interviews benefit from being able to provide an in-depth understanding of an individual's experience that would be difficult to obtain from traditional self-report measures. Despite this advantage, phenomenological interviews are difficult and time consuming to analyze, not standard across participants, and are prone to the same limitations of other subjective self-report methods. Overall, phenomenological interviews provide rich first-person data but are limited due to the difficulty of analysis and subjectivity of participant responses.

Each methodology contributes to the understanding of various experiential states, but the individual impact of these contributions is restricted due to the nature and limitations of each approach. One approach to overcoming the limitations of each method is to integrate the use of all three in interdisciplinary studies. Neurophenomenology is a promising approach based on this kind of integration (Varela 1996). Our project subscribed to the neurophenomenological method because it is an integrated and holistic approach for understanding and exploring experience (but see Chapter 4 concerning the limitations of this method).

2.1 Participants

Thirty-eight participants (13 males, 25 females) with an average age of 20.94 years were recruited from general psychology classes at the University of Central Florida. Participants were provided with an informed consent form and restrictions list. Participants were screened on the following criteria: age, right-handedness, color blindness, seizures, tobacco or caffeine intake, alcohol consumption, and ingestion of antipsychotics or antidepressants. Most of these restrictions are also requirements for astronauts preparing for space travel. The full course of the experiment lasted 2.5 hours.

2.2 Narrative

As each participant prepared for the experiment, a space-flight narrative was initiated, explaining that he or she would be involved in a simulated space flight. The experimenters communicated the following scripts: "Welcome to your pre-flight preparation. Before we begin, do you need to use the restroom? Let's begin the first phase by having you follow me to the cockpit where we will begin your astronaut qualification examination." Various events outlined in the narrative included connecting the participants to the physiological instruments and completion of a short battery of questionnaires.

The narrative continued as the participant "suited up," i.e. as the participant was connected to the various physiological sensors with a detailed explanation of the devices.

2.3 Questionnaires

After suiting up, a 5-minute resting baseline was initiated, requiring the participant to remain still and quiet while gazing at the blank monitor. Then, a short battery of online questionnaires was administered in the VSL, allowing the participants to get acclimatized to the new environment. The questionnaires included the following:

- *Need for Cognition Scale (NCS)* (Cacioppo and Petty 1982) was used to determine the tendency to want to be engaged in active thinking and complex problem solving.
- *Multiple Stimulus Types Ambiguity Tolerance* (Mclain 1993) was used to assess openness and acceptance of items or concepts that are not concrete.
- *Tellegen Absorption Scale (TAS)* (Tellegen and Atkinson 1974) was used to understand likelihood to be immersed and enveloped in an environment.

- *Simulator Sickness Questionnaire (SSQ)* (Kennedy et al. 1993) was used to ensure that participants were at full capacity in the simulator and not negatively impacted by the simulator. The SSQ was not for use in formal analyses of results, only for monitoring the participants' state in the simulator during the experimental session.

2.4 Experimental design

When participants were settled and seated alone in the VSL, we began the countdown and launch sequence. Following the launch, one portal opened, presenting one of the four space views. Each one lasted for 12 minutes. The experiment was a 2 (earth or deep space) x 2 (object or no object) mixed design with repeated measures on the first variable. Each participant received two experimental conditions in a counter-balanced order. Participants completed the SSQ on the computer monitor following each condition.

The experimental design was intended to enable investigation of experiencers of AWCH compared to non-experiencers of AWCH by maximizing the opportunity to induce those experiences with different space views that map to those reported by astronauts.

2.5 Physiological measures

Each participant was connected to electroencephalography (EEG), functional near-infrared spectroscopy (fNIR), and electrocardiogram (ECG) equipment.

- EEG and ECG: The Advanced Brain Monitoring X10 EEG/ECG system measured electrical activity of the brain and heart in response to the above-described stimuli. Nine channels assess neural activity yielding the power spectral density outputs of alpha, beta, and theta waves for each channel. These channels combine for measures of right and left hemisphere responses. The remaining 10th channel is dedicated to receiving ECG input from the electrical activity of the heart. Measures derived include heart rate (HR), HR variability, and interbeat interval (IBI).
- fNIR: The Covidien INVOS-5100C Cerebral/Somatic Oximeter System sends near-infrared rays into the brain with one sensor and receives metabolic usage at another sensor, measuring oxygenation of cortical tissue in the pre-frontal cortex.

PhysioSync, a computer software program developed at the Institute for Simulation and Training, allowed for synchronized multisensory recording and precise correlations between EEG, ECG, and fNIR and the simulated environment.

2.6 Phenomenological interviews

Following the landing sequence, participants were disconnected from the equipment and led to an interview room for "debriefing." Trained personnel conducted a 30-minute interview using a specific phenomenological interview technique. Interviews were audio recorded and then transcribed.

The interviews were based on a method developed in the Varela lab in Paris (Petitmengin 2006). The approach is characterized by open questions. That is, the interviewer simply asks the participant to talk about his or her experiences during the launch and visual simulations in the VSL, without providing a pre-determined vocabulary or list of items to be discussed. The interview explores the participants' experiences during events in the VSL (launch sequence, first portal opening, questionnaire, second portal opening, and landing sequence). Subsequently, the interview is remapped onto the simulation timeline, lining up the interview description with the sequence of events in the simulation.

The focus of the interview was on the participant's (first-person) lived experiences, that is, what they were experiencing, feeling, or thinking about as they viewed the space scenarios through the portholes (see sample passages from interviews, below). In cases where participants started to digress from experiential descriptions (e.g., where they started to talk about theories or non-experiential artifacts), interviewers were trained to lead the participants back to their experiences in the VSL. Interviewers also periodically paraphrased and repeated the participants' statements, paying careful attention to use the participants' own words, in order to confirm that the interviewer understood. Such confirming reviews also provided the participant with opportunities for adding something they forgot or for clarifying their descriptions.

The interviews were completed by trained personnel other than the experimenter who worked with the subject prior to and after the simulation. Therefore, for the interviews, there was a hand-off from one researcher to the interviewer, and a change of environment. The interviews were conducted in a separate room. The interview was cast as a post-flight debriefing by a member of "ground control," which was consistent with the space-flight narrative established at the beginning.

The transcriptions of the phenomenological interviews then underwent hermeneutical analysis (by primary and secondary interpreters) using the consensus categories (described in Chapter 2). The question was whether experience in the simulations replicated the categories found in the astronauts' experience.

2.7 Follow-up questionnaire

At the end of the interview process, we administered the Brief Multidimensional Measure of Religiousness/Spirituality (BMMRS) (Idler et al. 2003; Masters 2013; Masters et al. 2009). The BMMRS is a 34-item measure that evaluates religiousness/spirituality in seven areas: Experiential Comforting Faith, Negative Religious Interaction, Personal Spirituality, Punishing God, Religious Community Support, Private Religious Practices, and Forgiveness. The BMMRS is a reverse scored survey. The categorical questions included items where the participant is asked to agree/disagree with statements about private religious practices (e.g. "Besides religious services, how often do you take part in other activities of a religious nature?") Higher scores on this question, for example, indicate lower levels of private religious practices.

3 Results of the first experiment

Our approach to data analysis was to first determine whether the experimental design elicited experiences similar to those reported in the astronauts' texts. If categories derived from those texts could be found instantiated in the phenomenological interviews, then a tally of those who experienced awe and wonder (as defined by the consensus categories) would ensue for each of the four experimental conditions. Then, data from the questionnaires and from the physiological and neurophysiological measures would be correlated to the phenomenology of experiencers and non-experiencers of awe and wonder under each condition.

3.1 Hermeneutical analysis

Results from the hermeneutical analysis of the phenomenological interviews suggest that the participants' responses to the VSL simulations are similar to those expressed in the astronauts' reports. Here we provide some indications of this replication by comparing an analysis of the phenomenological interviews with the analysis of the original astronauts' reports.

First, consider two reports from astronauts indicating aesthetic appreciation, scale effects, and a perspectival (moral) shift, and then compare them to two interviews that touch on the same experiences. As indicated in Chapter 2, scale effects are indicative of experiences of awe; perspectival (moral) shifts reflect experiences of wonder.

Astronaut report 6. (Greg Chamitoff's in-flight journal, 11/15/08)

From these windows, the Earth is so obviously floating in an endless void, and the feeling that washes over you is the sense of scale of the universe. The feeling I got was one of recognition that we are living on a such a tiny island in a vast ocean [**scale effects**].

Astronaut report 24. (Marc Garneau. Interview cited in White 1987, p. 249ff)

One is just the sheer wonder of looking down at the Earth. It is very, very beautiful... particularly when you see it interface on the edge with space [**aesthetic appreciation**]. There you suddenly get the feeling that, hey, this is just one small planet which is lost in the middle of space... [A] very important feeling about the fact that we're just drifting through an immense universe... [**scale effects**]. [Y]ou become a little more conscious about the fact that we shouldn't be doing silly things on Earth like fighting and killing each other [**perspectival (moral) shift**].

Compare these to the descriptions offered by the participants in the phenomenological interviews that followed the simulations.

Participant 3

I guess I was a little... not shocked but amazed a little bit cause I find it fascinating—outer space, space travel, things like that. So, it's a view that you don't see regularly. So, it kind of got me interested in seeing the difference between what you see every day about Earth, and you don't really look at it from such a big point of view that everybody's on that small little planet, and you're so far away now. So, it's a different view on Earth.... I think it just really makes you feel less important when you look at everything in such a view like that. You're just a speck on the Earth that's in a universe of many different planets. You're small compared to everything else [**scale effects**], and I didn't feel too bad, but it kinda makes me feel like my problems now are not really as big as I think they are compared to everything else in the world [**perspectival (moral) shift**].

Participant 1

I was just kind of looking at the stars and admiring what I saw. I think it's the vastness of reality.... then I start thinking of how huge our universe is. Like, just looking at this, and this is just a little part of what I'm looking at and how much more there is [**scale effects**].

That's the part that I admired...The beauty of the lights and all that [**aesthetic appreciation**]; but, to me, somebody created all that. That blows me away.... [My] mood would be like: taken back, in awe, I was definitely admiring, definitely peaceful, relaxed, and then just like in awe of my mind taking me where I was really reflecting on how huge space is.

In one case, during much of the interview, one of the participants indicated that he did not have sufficient sleep the night before, and that he found himself bored and often sleepy (his eyes starting to close) during the visuals presented in the VSL. Even this participant, however, expressed something similar about scale.

Participant 5

It was just like...It is a vast amount of space that we are not going to be able to, as a species of homo sapiens, we're not going to be able to identify and figure out everything.... Even though we try to, there's just too much going on and the Earth is always gonna have its secrets.... You feel so small compared to everything else. So, you're looking at this vast amount of space...where your home is barely the tip of the needle point.... How big are you compared to that? Like, you're probably only maybe an eighth, maybe a sixteenth of that needle point. It makes you kind of realize that there's a lot more out there.... It's not just the little space that you live and work and breath in [**scale effects**].

Consider the following examples, more reflectively religious in tenor; the first from an astronaut's report, the second from one of the participant interviews.

Astronaut report 28. (Bill Nelson, interview cited by White, 1987, pp. 285ff)

I frequently recalled what King David had written thousands of years ago in Psalm 19: "the heavens declare the glory of God, and the firmament showeth his handiwork". That order, that creation was very apparent to me as I looked back on this beautiful planet that looked so fragile at the same time.

Participant 1

I did reflect on a scripture that talks about God holding the universe in his hand. So as I kept going and going and going [through the

> simulation] and I thought, wow that's pretty cool, that here it is, just this short bit of time that I see all of this and how big it must be that he holds it in the palm of his hand.... Yea, basically I was just at peace and knowing that I could never comprehend truly the vastness of it all....I just kind of reflected on how cool it was to be able to see how small Earth is from my perspective up in space. You know, we think of the Earth as huge, but when you're looking down at it you can start seeing how from there it could almost fit in the palm of your hand, then it was like, kinda cool.

These two reports clearly have a religious tenor; Participant 1 also clearly expresses experiences of scale effects.

Table 3.1 provides some comparative figures between astronaut reports and the phenomenological reports from participants in experiment 1.

3.2 Questionnaires

We were able to identify a derivative variable through further analysis of differences between experiencers and non-experiencers of awe and wonder in light of questionnaire responses. In particular, we conducted correlations for each of the questionnaires with awe categories and then wonder categories for each experimental condition. It is important to note that the texts of the interviews and the astronauts' reports were the central focus for determining if participants had similar experiences to those reported by astronauts. Since astronauts did not complete the questionnaires or wear the physiological sensors we employed in our study, these cannot be directly compared to the astronauts' responses. However, all of this data may provide interesting insight as to why only some participants and, by extrapolation, why only some astronauts, have awe and wonder experiences, whereas others do not.

In the category of experiencers of awe during *(1) the earth condition*, there were positive correlations between the experiential comforting faith subscale ($r = .525$, $p = .021$) and the private religious practices subscale ($r = .506$, $p = .027$) on the BMMRS. The BMMRS is reverse scored, which means that participants who experienced awe showed lower levels of experiential comforting faith and private religious practices. There were no significant correlations in experiencers of wonder during the Earth condition.

For experiencers of awe during *(2) the deep space condition,* there is a positive correlation between the sensory-perceptual absorption scale ($r = .481$, $p = .037$) on the TAS, and a negative correlation between the private religious practices subscale ($r = -.495$, $p = .031$) on the BMMRS. Again, since the BMMRS is reverse scored, this correlation suggests that

Table 3.1 Frequency of categories expressed by astronauts and participants in Experiment 1

Consensus categories	Frequency in astronaut reports (51 reports, c. 23,000 words)	Frequency in post-simulation interviews (19 interviews; c. 37,500 words)
Aesthetic appreciation	17	14
Captured by view/ drawn to phenomenon	6	3
Change (internal, physical)	3	0
Connectedness, feelings of	2	8
Contentment	1	22
Interest/inquisitiveness	7	22
Disorientation	3	8
Dream-like	2	4
Elation	2	0
Emotion (general)	3	2
Experience-hungry (wanting more, setting up to have experience)	4	6
Floating (bodily – related to weightlessness)	6	1
Floating in void (not related to weightlessness)	4	6
Fulfillment	2	2
Home, feeling of being at...	3	8
Inspired [reflective]	3	18
Intellectual appreciation	4	7
Joy	3	3
Nostalgia	3	5
Overwhelmed	11	7
Perspectival change (spatial)	10	16
Perspectival shift (moral, internal)	4	42
Peace	3	6
Pleasure	3	3
Poetic expression	9	1
Responsibility (towards others)	3	4
Significant sensory experience (visual, silence)	6	21
Sensory overload	7	11
Surprise	7	2
Unity with (feeling of oneness with; holistic feeling)	8	0
Unity of external (the earth, universe, people on earth)	7	1
Sublime	1	1
Totality (wholeness of what is experienced)	1	6
Scale effects (feelings of vastness of universe; feeling of smallness/ insignificance within the vast)	7	22

for participants who experienced awe, levels of private religious practices were lower.

In the category of experiencers of awe during *(1) the earth condition,* there were positive correlations between the experiential comforting faith subscale ($r = .525$, $p = .021$) and the private religious practices subscale ($r = .506$, $p = .027$) on the BMMRS. Since the BMMRS is reverse scored, this indicates that participants who experienced awe showed lower levels of experiential comforting faith and private religious practices. There were no significant correlations in experiencers of wonder during the earth condition.

For experiencers of awe during *(2) the deep space condition,* there is a positive correlation between the sensory-perceptual absorption scale ($r = .481$, $p = .037$) on the TAS, and a negative correlation between the private religious practices subscale ($r = -.495$, $p = .031$) on the BMMRS. Again, since the BMMRS is reverse scored, this correlation suggests that for participants who experienced awe, levels of private religious practices were lower.

For experiencers of awe during *(3) the earth-with-object condition,* there were negative correlations between the intuition ($r = -.488$, $p = .016$) and trance ($r = -.512$, $p = .011$) subscales on the TAS, while a positive correlation was found for the need for cognition (on the NCS) ($r = .512$, $p = .010$). No significant correlations were found for experiencers of wonder during the earth-with-object condition.

For experiencers of awe during *(4) the deep space-with-object condition,* there was a negative correlation with the private religious practices subscale ($r = -.551$, $p = .005$) on the BMMRS. This correlation again shows that for participants who experienced awe, their levels of private religious practices were lower. There were no significant correlations for experiencers of wonder during the deep space-with-object condition.

3.3 Physiological measures

Looking at the relation between the physiological and neurophysiological responses and the experiences of awe and wonder, participants exhibited negative correlations in *(1) the earth condition* between experiences of awe and levels of left hemisphere theta ($r = -.478$, $p = .038$) and right hemisphere theta ($r = -.466$, $p = .044$). Experiencers of wonder during the earth condition resulted in no significant correlations.

Experiencers of wonder during *(2) the deep-space condition* exhibited a positive correlation for average IBI ($r = .469$, $p = .043$). There were no significant correlations with awe experiencers.

There were no significant correlations found in *(3) the earth-with-object* or *(4) deep space with object* conditions between experiencers of awe and wonder and the physiological measures.

4 Discussion

The primary aim of this study was to replicate and explain (in psychological, physiological, and phenomenological terms) certain kinds of experiences had by astronauts during space flight. A secondary aim was to explore a specific methodology that employed simulation technology and phenomenological report in a neurophenomenological design. As we'll make clear in the following chapter, we learned as much about making this method work properly as we did about experiences of awe and wonder. In the end, we treated this experiment as a pilot study on which we based our second experiment.

Not every participant had experiences that could be classified using the consensus categories, or in the broader terms of aesthetics, spirituality, or religiosity. Nor, as far as we know, did every astronaut or cosmonaut. The results, however, supported the primary aim for the present study, replicating, within a simulated environment, experiences previously had by individuals only in the special environment of extraterrestrial space.[1] Experiential descriptions given by participants in this first experiment resemble those made by astronauts in their journals. From the hermeneutical analysis of the interviews, it appears that many, even if not all of the relevant astronaut experiences were replicated. For example, no experiences of *elation* or of *unity with the universe* (feeling of oneness with everything) were generated in the simulations. The most frequent experiences were:

- perspectival (moral) shift (42 instances);
- interest/inquisitiveness (22 instances);
- contentment (22 instances);
- significant sensory experience (e.g. impressed by silence or visuals) (21 instances);
- inspired (reflective experience) (18 instances).

The phenomenological interviews allowed us to determine high-level classifications of *experiencers* versus *non-experiencers* of awe and wonder,

[1] We understand experiences to be individuated not only by their phenomenal character but also by the aspect of intentionality, i.e., what they are about or what object is being experienced. Accordingly, it may be quite possible to experience awe while standing in a dessert, or on a mountain top, or in front of a piece of art, but such awe experiences are differentiated, at the very least, in their intentional aspect. Whether awe is phenomenally the same or not in each case is a further question. The phenomenality of the experience may be closely connected and modulated by intentionality (see Gallagher and Aguda 2015).

enabling correlational analyses with data from physiological and neurophysiological measures, and the questionnaires. Consider, first, some significant findings from the questionnaires.

The most surprising finding coming from the questionnaires concerns the BMMRS, which measures levels of religiosity. On the BMMRS, the *experiential comforting faith* subscale is comprised of items that reflect an experience of a close relationship with a higher power by asking for responses to such statements as "I feel God's presence" and "I desire to be closer or in union with God." Additionally, the *private religious practice subscales* consist of items that identify religious/spiritual behaviors in terms of responses to questions such as "How often do you meditate?" and "How often do you read sacred religious texts?" Experiencers of awe in the Earth condition indicated lower levels of experiential comforting faith and private religious practices compared to non-experiencers of awe. The results indicate that participants who experienced awe in the earth condition tended to have a weaker perceived relation with a higher power and performed less religious/spiritual behaviors in their everyday life. A weaker finding contrasted with this. Participants who experienced awe during the *deep space-with-object* condition tended to engage in religious practices more often than non-experiencers of awe. Taken together, these results show generally that religious/spiritual traits are factors for some, but not all people who experience awe. These were clearly results that needed further investigation.

The intuition subscale on the *Tellegen Absorption Scale* (TAS) measures a person's ability to "feel" and "experience" the world around him or her without explicit or studied perception. They are able to get a gist of the situation without reflectively sizing it up. For example, the subscale asks for responses to the following statements: "At times I somehow feel the presence of someone who is not physically there," and "I often know what someone is going to say before he or she says it." Additionally, the trance subscale measures mystical experiences through responses to items such as "I think I really know what some people mean when they talk about mystical experiences," and "Sometimes I experience things as if they were doubly real." Participants who experienced awe during the *earth-with-object condition* tended to have lower scores in the category of intuition and trance as measured by the TAS, than those who experienced awe in other conditions, or those who did not experience awe in any condition. These findings may reflect the possibility that the presence of the ISS in the simulated visual allowed participants to grasp the scale of the earth with better perceptual precision than would be possible without a perspective-producing or reference object. This interpretation assumes that gaining a good sense of the earth's size may be connected with the experience of awe in this circumstance. This suggests that for people who are not intuitive thinkers, simulations designed to provide concrete images are better for inducing awe. This, again, is something that would require more study.

Related to this last finding, experiencers of awe in the earth-with-object condition tended to have higher need for cognition scores (which can be interpreted as a measure of preference to think critically and to apply logic to derive explanations) as measured by the NCS. This result would reinforce the hypothesis that the appearance of the ISS in the visual field may provide a definite point of reference facilitating a clarification of the earth's size – assuming again that getting a good sense of the earth's size may motivate the experience of awe.

Experiencers of awe in the deep-space condition rated higher on the sensory-perceptual absorption scale on the TAS. Questions on this scale ask about a person's sense of being connected with, or separated from, what is physically present. In other words, someone might be so immersed in a task that their consciousness feels changed; or their engagement with physical objects or with nature sends them into thoughts of something sentimental like a favorite piece of music. This suggests that individuals who experienced awe in the deep-space condition likely "got lost" in the vastness of the stars. This has implications for designing simulations to induce that feeling of overwhelming depth.

There were also significant findings in physiological and neurophysiological data for experiencers of awe and wonder. Experiencers of wonder showed higher IBI, measuring the amount of time between heartbeats, in the deep-space condition. It is possible that those struck with wonder in the deep-space condition were holding their breath while pondering the universe, leading to longer durations between heartbeats. It also might be that individuals who experienced wonder in this condition were in a more meditative state leading to longer intervals between heartbeats. We did not compare the likelihood of participants to experience awe and wonder together because we were looking at these as distinct phenomenon. It is possible, however, that a person who experiences awe is more likely to experience wonder. If that is the case, then the above-mentioned correlations between the experience of awe in the deep-space condition and higher ratings on the sensory-perceptual absorption scale would add support to this IBI finding of such participants being fully immersed in or entranced by the view. Also, it's an open question whether the phenomenological interview could be more fine tuned to explore such physiological changes, since the participant in the simulation may not be aware of holding his or her breath or being in a meditative state. There was no indication of such experiences in the phenomenological interviews connected with the first experiment.

Our results showed that several EEG metrics were able to differentiate between AW experiencers and AW non-experiencers and showed significant differences in several traits, as measured by the BMMRS. Additionally, both phenomenological interviews and EEG beta differences indicated that the Earth view was more influential in eliciting AW statements compared to the Deep Space view.

In order to identify physiological markers that correlate with experiences of awe and wonder, EEG, ECG, and fNIR metrics were compared among AW experiencers and non-experiencers. Increased measures of left hemisphere (LH) and right hemisphere (RH) theta were found among non-experiencers of awe compared to experiencers of awe during the Earth condition. Traditionally, increases in theta have been interpreted in two contrasting ways (Paus and Zatorre 1997); increases in frontal lobe theta have been interpreted as representing enhanced cognitive activity/working memory load (Smith et al. 2001; Gevins et al. 1997), whereas increases in wide-spread scalp theta have been interpreted as representing increased drowsiness and fatigue (Paus and Zatorre 1997). The differences found in the present study, across LH and RH, and not just frontal lobes, most likely reflect the latter interpretation. AW non-experiencers during the Earth condition may have experienced greater levels of drowsiness or a decreased level of alertness compared to experiencers of awe. According to Schacter (1977), participants who experience a drowsiness-related theta response also have a decreased awareness and ability to actively interact with the environment they are in. However, in order for an AW experience to occur, one's attention must be directed toward the stimuli (Shiota et al. 2007). As a result, the widespread theta response can be used to differentiate AW non-experiencers, whose attention is not sufficiently directed toward the stimuli, from neutral and AW experiencers who are alert and attentive to the stimuli.

This finding is also supported by the interviews of the AW experiencers compared to the non-experiencers, with the non-experiencers reporting boredom and inattention in regard to the stimuli. During the interviews, AW experiencers also mention feelings of boredom and inattention but, on average, not until much later in the simulation. This later distraction or boredom reporting by the AW experiencers is likely associated with the intense theta changes occurring in the physiological measures around the 8–10 minimum period for time spent viewing the Earth simulation. These theta changes from resting baseline during the later period resemble the physiological changes that occur during a vigilance task. Operators in vigilance tasks are most frequently highly motivated individuals who find it difficult to maintain attention to the task over time, increasing workload (Reinerman-Jones et al. 2010). For this reason, we reduced the time on task from 12 minutes in Experiment 1 to 7 minutes in Experiment 2.

The correlation data provide insight into which experimental conditions are associated with individual differences. However, it does seem that these differences are more strongly related to awe and offer less insight into what differentiates experiencers and non-experiencers of wonder. Furthermore, it seems that this trend is more relevant to the earth conditions than to the deep-space conditions. This again suggests considerations about simulation design, based on those individual differences, and we decided that further research would be needed to more directly assess those design hypotheses.

The results of this experiment confirmed that the general methodology followed in this study is a viable one and demonstrated the promise of simulation technology for designing experiments in psychology and cognitive science. The present experiment enabled replication for average, untrained participants, in a simulated environment on Earth. of experiences related to space travel first reported by astronauts, who are among an elite few who have actually had the opportunity to see the earth and space from an extraterrestrial vantage. This supports the use of carefully designed and developed simulation technology in the scientific investigation of such experiences.

Successful induction of awe and wonder experiences were enabled by a selection of realistic images (images actually taken in space), but embedded in a dynamic simulation and surrounding environment that helped the person suspend disbelief about being in space. This study recommends the use of expert analysis of imagery for incorporation into simulation. Studying phenomena that occur naturally and are challenging to capture in the laboratory require careful review of whatever is available to replicate the environment in simulation.

Furthermore, attention to detail described in reports of the experiential phenomenon, such as the physical context (i.e., whether the person was floating or sitting at a workstation near a window) as well as the phenomenological context (i.e., whether the person was looking out the window just after working really hard so that his physiological responses were on the intense side or a relaxing occasion) is important.

The mixed-reality environment (including physical structure plus simulated visuals) allowed for the induction of awe and wonder experiences. There are, however, several limitations involved in both the choice of this kind of simulated environment and the specific experimental design that was used. These limitations (discussed in the next chapter) motivated a redesign of this study.

4

Neurophenomenology and Simulation: Philosophical Ground Control and a Sharpening of Our Tools

As research into emotions and affective phenomena, such as awe and wonder grows, so will the methodological challenges for addressing such inquiries. These inherently nebulous and conceptually ambiguous topics draw the interests of researchers from various disciplines, each coming with distinct perspectives and biases concerning what counts as good answers to their questions. While this meeting of perspectives will inevitably introduce some methodological conflict, there are specific practical considerations that can assist in disentangling such conflicts.

This chapter focuses on how to perform the science when the topic requires that a number of disciplines be accommodated. We suggest that this is going to be the case whenever the topic concerns human experience and behavior. That is, in studying things like emotions, consciousness, aesthetic experience, spirituality, or the nature of self, to get a full account (or at least a fuller account than can be delivered by one discipline alone), one needs to work across disciplines. In an effort to support that goal, this chapter offers some basic considerations. We review the "lessons learned" from Experiment 1. In that experiment, we attempted to create a methodological design that is both non-reductionist and disciplinarily inclusive. The first section of this chapter positions this task within the context of neurophenomenological approaches. Next, we consider some philosophical objections to neurophenomenology (NP). Two experimental designs are then contrasted by applying three key methodological lessons. The chapter closes by considering the lessons learned from Experiment 1 and the place of NP within a larger discussion of data collection and interpretation.

1 Triangulation and charting a methodological course

Each discipline brings a unique set of tools to this project. Typically, researchers can apply a range of tools and techniques that fit neatly into their own disciplinary space. Yet, different disciplines have different starting points and make different assumptions about their topics of study, and these inform what does or does not go into their different collections of tools. Methodological clarity comes, in part, from a discipline's identity, shaped by its starting points and assumptions regarding what it does or does not do.

When disciplines come together to engage in truly interdisciplinary acts of science, the methodological imperatives can become far less clear. As one may suspect from the name alone, neurophenomenology (NP) is a combination of phenomenology and neuroscience, or the study of experiential phenomena through a neurobiological perspective. NP as it is used in this volume, applies to an understanding of *how* consciousness can be examined. It combines methodologies from several disciplines into a new set of tools that can triangulate on the kind of experiences we want to study.

The notion of triangulation derives from the use of stars for navigation and can be defined in geometrical terms. It allows the navigator to determine the location of a point (specifically where the navigator is) by measuring angles. That is, you can specify your location on Earth by using a fixed baseline defined by two stars (by measuring the angles from your location to those stars). The Greek philosopher, Thales, who we mentioned in the first chapter, used triangulation to estimate the height of the pyramids (measuring the lengths of his own shadow and of the pyramid's shadow at the same time, and then calculating using the ratios to his height).

The concept of triangulation has been used in cognitive science to indicate a combination of methods designed to intersect on a target subject matter. Owen Flanagan (1992), for example, suggested that to study consciousness one needed to combine psychology, neuroscience, and phenomenology – a combination reflected precisely in the neurophenomenological program as proposed by Francisco Varela (1996), and described in Chapter 1. One might consider the following objection to this combination of methods; however, deriving from an observation made by the philosopher, Antony Flew (1966) concerning the idea of

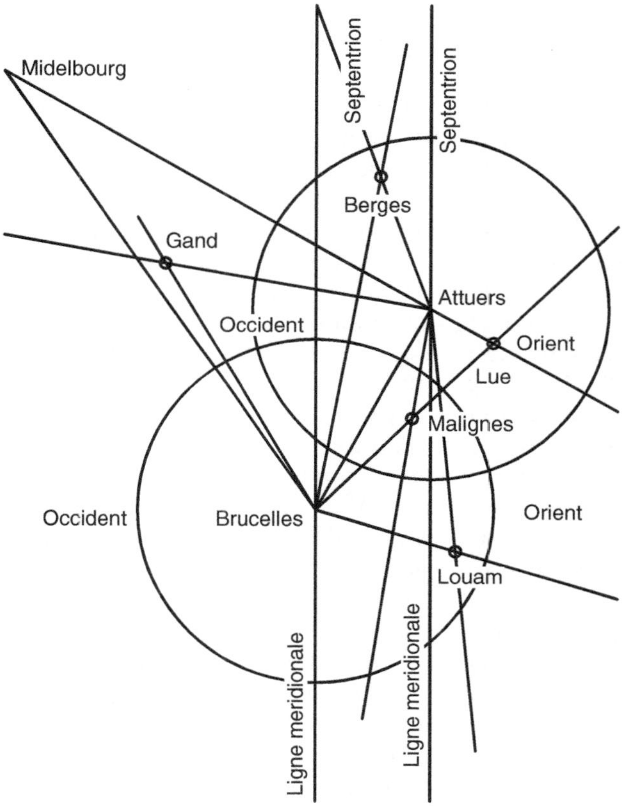

Figure 4.1 Gemma Frisius's 1533 diagram introducing the idea of triangulation into the science of surveying. *Libellus de locorum de scribendorum ratione.* [Image in the public domain, source: http://www.math.yorku.ca /SCS/Gallery/images/G-F_triangulation.jpg].

using many arguments for the existence of God. He suggested that this would be like using a series of leaky buckets. Putting all the leaky buckets together doesn't mean that together they will hold water. Accordingly, one possible objection to triangulation is that the limitations of each method remain, and the combination of methods doesn't really deal with those limitations. The response to this kind of objection is to look at the correlations between the results from the different methods instead of using them to shore up each other. Part of our method, then, involves using a kind of second-order analysis. We first generate data in various ways and then pay attention to the similarities and differences in those results. So, our approach to triangulation is less like collecting water in leaky buckets and more like gathering grapes from different vineyards, tasting and comparing, and possibly blending to attain the best vintage. *In vino veritas*. As in any methodological approach, as it is used and as it matures, it attains more clarification.

As we indicated, the notion of phenomenology, as a systematic study of first-person experience, derives from the philosophical work of Husserl and others in the first part of the twentieth century. In the 1990s, Varela and his colleagues introduced the concept of a *naturalized* phenomenology, which brought these philosophical perspectives on first-person experience into the natural scientific laboratory (Varela 1996; Petitot et al. 1999). Varela's program of research combined neuroscience, using EEG and brain scans to study experience as reported by subjects trained in phenomenological reflection.

For example, an experiment conducted by the Varela group in Paris (Lutz, Lachaux, Martinerie, & Varela 2002) demonstrated how subjects trained in phenomenological method could participate in psycho-physical experiments on perception. Lutz et al. studied subjective parameters concerning attention or readiness for task (vs. being distracted or unready), spontaneous thought processes and strategy decisions that occur in many empirical testing situations that target specified cognitive tasks. Their experiment helped to specify why successive brain responses to repeated and identical stimulations, recorded for example by electroencephalography (EEG), are highly variable. Lutz et al. were able to correlate phenomenological reports (by trained subjects) with behavioral reaction times and dynamical brain activity, recorded by EEG. The phenomenological training undergone by the subjects included pre-trial development of a shorthand vocabulary for indicating, for example, differences in attentional readiness. Using these categories of attention, subjects were then able to easily report on the attentional aspect of their experience during trials involving 3D perceptual illusions (Figure 4.2). The reports during the main trials revealed subtle changes in the subject's experience due to the presence of specific differences in attention or cognitive strategy. The first-person data correlated with both behavioral measures (reaction times) and dynamic descriptions of the transient patterns of local and long-distance synchrony occurring between oscillating neural populations, specified as dynamic neural signatures (DNS).

While attention and brain activity are classic topics in cognitive psychology and neuroscience, the Lutz et al. study included two distinct methodological features that are still uncommon practice in basic research.

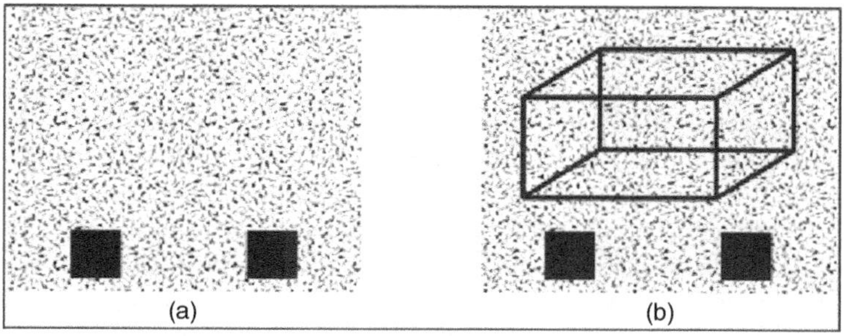

Figure 4.2 A 3D perceptual illusion. "Subjects are shown random-dot static images on a digital monitor, and at an auditory signal, they are asked to fuse two squares at the bottom of the screen (a), and to remain in this eye position for seven seconds. The random-dot pattern was then changed to a slightly different random-dot pattern with binocular disparities (an autostereogram). Subjects were then able to see the 3D illusory geometric shape (represented in b). They were instructed to press a button with their right hand as soon as the shape had completely emerged. The subjects then gave a brief verbal report of their experience" (Lutz et al. 2002; Image from Lutz et al. 2002, with permission).

First, the subjects were extensively trained in phenomenological method. Their reports of experience (and the experimenters' interpretation of their reports) followed the three methodological steps outlined in Chapter 1.

1. *Suspending beliefs and theories regarding experience* (the *epoché*, or bracketing). Both experimenters and participants strive to avoid any tendency and impulse to indulge in analyzing or judging their experiences. The goal is to report the experience while avoiding bias from within or without.
2. *Gaining reflective intimacy with the experience* (the phenomenological reduction): To generate accurate accounts, participants must be capable of the self-reflection that can provide an intimate (first-person) account of the experience in connection with exposure to a given stimulus, focusing more on *how* I, as subject, am experiencing something rather than on the *what* I am experiencing.
3. *Seeking intersubjective validation*. While first-person subjective accounts are essential for the process, NP demands rigorous attention to the handling of this form of data in verifying the phenomenological description and helping to develop a shared vocabulary for such experiences, for example, by comparing many different descriptions of a particular experience, and elucidating differences and similarities or invariant features.

Second, and following from this third step, their analysis bundled experiences into "phenomenological clusters" – that is, into a set of categories of experience. The concept of phenomenological clusters is similar to the concept of consensus categories defined in Chapter 2. These clusters were generated from the phenomenological reports that followed initial (pre-) testing, and they provided reportable categories that were used in the main trials. In contrast to traditional neuroscience experiments where the *explanandum* (the experience to be explained) is predefined in general folk psychological terms (the subject is presumed to simply "attend" to the stimulus), the neurophenomenological approach allowed the subjects to define their own experience on the basis of nothing other than their own experience. Nothing was predetermined about what that experience would be like. The subjects' own experiences generated the categories for subsequent testing with these same subjects, allowing the experimenters to identify more precise correlations between the electroencephalographic (EEG) data and the experiential accounts. The result was the identification of dynamical neurological correlates to experience, correlations that offered a more refined understanding of attention. Factors that would otherwise be dismissed as "noise" in other neuroscientific studies were shown to be explanatorily relevant in the neurophenomenological methodology. The consideration of first-person experiential data in these experiments allowed human performance and neurobiological data to be integrated into a fuller picture than would be captured with traditional neuroscience and psychology.

NP research, then, executes these steps by avoiding pre-defined categories; by adhering to the subject's experience as a genuine source of data; by avoiding the imposition of external biases that would adhere to assumptions about what an experience should be like. Open-ended questions (not unlike those frequently employed in clinical psychology) are used in the neurophenomenological process to support participant reflection while avoiding the imposition of biases and judgments.

Outside of the laboratory, however, or even in controlled experiments where the experimenters have limited time with participants, it can be difficult for the participants to genuinely examine, let alone articulate, their experiences. Phenomenological training is not possible in all experimental circumstances. Without training, researchers cannot assume that a participant is self-consciously aware of her experience, and even if she is aware in an introspective fashion, she may not be able to reflect upon and describe that experience in a phenomenologically rigorous manner. In such circumstances, the burden falls to the researchers to use techniques to support the attempt to get at the experiences and to help participants become aware in a way that remains descriptive and non-judgmental.

This is where the phenomenological interview comes into play. Working with epileptic patients in Paris, the Varela group developed a phenomenological interview technique that can substitute for training experimental subjects (Petitmengin 2006; 2010). The interview technique allows experimenters to access first-person descriptions from subjects who are not trained in phenomenological method. The interviewer, who is trained in phenomenological interview techniques, helps the subjects to reflectively return to their experiences and to explicate them in their own terms. The interviewers keep the subject's attention focused on the experiential aspects rather than on his or her opinions, beliefs, or theories about the experience. Across a sufficient number of subjects one can also attain intersubjective validations in this manner.

We've already described how this interview method was used in the first experiment (Chapter 3). We highlight the following steps.

1. Open-ended questions are key. The interviewer asks the participant to talk about his or her experiences during the launch and visual simulations in the VSL. There is no pre-determined vocabulary or list of items provided.
2. The interview explores the participants' experiences during events in the VSL. The focus of the interview is on the participants' (first person) lived experiences, that is, what they were experiencing, feeling, or thinking about as they viewed the space scenarios.
3. When participants digressed from experiential descriptions, the interviewers led them back to their experiences.
4. To confirm the descriptions that the participant has provided, interviewers periodically paraphrase and repeat the participants' statements, using the participants' own words. Participants are allowed to clarify or expand on the descriptions.

The Varela group employed a phenomenologically-informed open-ended questions technique to study epileptic patients awaiting brain surgery. The participants were in care for pharma-resistant temporal lobe focal seizure epilepsy, reporting a-symptomatic preictal states (Petitmengin 2006). That is, these participants were facing seizures that were not responding well to drug-based interventions that would occur without warning. Over the course of 18 months, Petitmengin's interviews with the untrained participants were used to create phenomenological clusters, which in turn were used to group EEG data. The results suggested that preictal experiences emerge far earlier than neuroscientists had expected, in some cases a week before the seizure (Petitmengin 2006).

Psychologists and neuroscientists often use questionnaires, which can transform qualitative data into quantitative data and thereby provide useful information about typical, and sometimes predictable, correlations between reported experience and neurological and psychological behaviors. However, as the studies described above suggest, NP can provide depth and higher precision to issues that are marked by the variation of human experience. That said, there are significant challenges that have rightly been raised against NP in respect to its empirical merit.

2 Enactivism and the hard science of consciousness

NP presents an opportunity for a methodological paradigm shift. While it is by no means the only approach to these kinds of interdisciplinary study, it is unique in its ability to triangulate on first-person experience, (including spiritual and aesthetic phenomena like awe and wonder), using a fully integrated interdisciplinary approach. NP, however, is open to criticisms, both about the way it addresses topics, and about the topics it chooses to address.

Criticisms of neurophenomenological approaches roughly come on two philosophical grounds that involve reductionistic science and an explanatory gap, respectively (Bockelman et al. 2013). The reductionist challenge asserts that explanations of consciousness and cognition ought to be reduced to purely physiological explanations if they are to be scientific, thereby eliminating the need to address what would be considered peripheral, subjective (and potentially erroneous) information collected in first-person reports. From the reductionist vantage point, we need to understand experience purely as a neurological phenomenon, elicited by (but separate from) the stimulus. The correct response to the reductionist challenge is to point out that there is a responsibility for science to "explain what there is" (Gallagher 2007, p. 311). What there *is* may not, necessarily be quantifiable, or entirely reducible without loss of some relevant aspects. The responsibility of science remains: if it exists, it ought to be explored. Experiences of awe and wonder do exist, and as experiences, they are first-person – that is, they are experienced from the perspective of the subject who is experiencing them. To offer an explanation of them in purely third-person, neurological terms seems to miss both the perspectival and phenomenal aspects of such experiences. Furthermore, the boundaries of experience are not necessarily drawn by neural connections; extra-neural factors also belong to the cognitive system.

Our study of awe and wonder takes this stance, considering the explanatory system to include not just the brain, but the brain-body-environment, where environment includes physical (or simulated), social, and cultural environments. The broader philosophical view that supports this NP approach is termed the *embodied/enactive* (or simply enactive) view (Varela et al. 1991), and is a specific version of the embodied cognition approach in philosophy of mind and cognitive science (see Chapter 8). The enactive view contends that a scientifically adequate account of experience requires that we understand, not only brain function, but also bodily responses in a physical, social, and cultural environment.

This enactive approach does not deny the utility of empirical analyses using traditional tools, but it does deny the reductionist desire to focus a zoom lens tightly on just the one aspect; typically, the neurophysiological aspect of cognition. The focus simply must be broadened, so that a fuller set of data can be included. The broader vision for neuroscience includes the incorporation of those things that may not be reducible to exclusively neural frameworks. NP resists the third-person objectification of what is rightly first-person subjective, and it values the unique source of data provided by an experiencer. Thus, scientists should not simply eliminate subjective experience and first-person accounts because they do not fit neatly into the rigid practices of most "hard science" (Bockelman et al. 2013). The real "hard" science, the difficult science, is the science that attempts to find ways to integrate first-person data with third-person data and to account for both neural and extra-neural factors.

A second critique is levied at NP's ability (or lack thereof) to address the "explanatory gap (Bayne 2004)." The challenge is to close the distance between physiological events and their counterparts in consciousness, thereby explaining (ideally with a tidy causal connection) how the chemical and electrical events within the brain bring about consciousness. Chalmers (1995) refers to this as the "hard problem" of consciousness, defining the explanatory gap as requiring a causal bridge between the brain and consciousness with all of its richly human features. This concern has significant merit from the perspectives of science and philosophy if one wants to avoid mind-body dualism. Scientific explanation is fundamentally causal explanation. In disciplines like sociology, economics, and psychology, researchers may indeed work with relationships between variables that are not explicitly causal, but finding a genuine causal link is still the aim of such sciences. With respect to establishing a causal theory of consciousness, closing the explanatory gap matters.

That said, it is not all that matters and it may not be as critical as it at first appears for the discussion of experience. To begin with, the call for closing the gap between consciousness and neurophysiology reflects a reductionist desire, and as we've just suggested, this may not be the most scientific approach to understanding consciousness. It follows, too, that the problem, as described, totally ignores the roles that non-neural factors, like bodies, affordances, interactions, and contexts, may play in the constitution of consciousness. Furthermore, the very definition of the problem reflects a dualism that contends that there is indeed a concrete distinction between consciousness and physiology, as if we have all agreed where one of these things ends and the next one begins. This suggests that the hard problem is ill posed. Although Varela (1996) framed neurophenomenology as an answer to the hard problem of consciousness, philosophically it may make more sense to challenge the legitimacy of posing the question in just this way.

Thompson, Lutz, and Cosmelli (2005), however, take the notion of the explanatory gap seriously. They point out that there are different dimensions to the explanatory gap, and distinguish between *conceptual, epistemological, and methodological* dimensions. Here's how they explain the distinctions.

- An adequate conceptual framework is still needed to account for phenomena that (i) have a first-person, subjective-experiential or phenomenal character; (ii) are (usually) reportable and describable (in humans); and (iii) are neurobiologically realized.
- The conscious subject plays an unavoidable epistemological role in characterizing the *explanadum* of consciousness through first-person descriptive reports. The experimentalist is then able to link first-person data and third-person data. Yet the generation of first-person data raises difficult epistemological issues about the relation of second-order awareness or meta-awareness to first-order experience (e.g., whether second-order attention to first-order experience inevitably affects the intentional content and/or phenomenal character of first-order experience).
- The need for first-person data also raises methodological issues (e.g., whether subjects should be naïve or phenomenologically trained).

All of these issues are clearly important ones that NP has to address, but we believe that these issues do not have to be framed in terms of the explanatory gap or the hard problem. Rather, we think these dimensions are generally and directly relevant to doing scientific investigations into experiences of all kinds, and they help to specify precisely the problems that have to be addressed. It is in this spirit that the work described next considers NP and its contribution to the study of experience. The aim of the following section is to refine the methods we used in Experiment 1, and more generally, make more precise our neurophenomenological methods. It charts methodological "lessons learned" during Experiment 1, and describes the methodological changes that were applied in our design for Experiment 2. These lessons focus on the design, execution, and data-collection phases of the basic experimentation.

3 Lessons learned

The questions that informed methodological choices for Experiment 1 were: (1) Is it possible to elicit awe and wonder experiences in a simulated space environment? (2) How can we rigorously examine these experiences from the NP perspective? The first question informed the design of the experimental simulation, whereas the second question informed the hypothesis testing from the simulation design to the metric selection to analytic techniques. We were convinced that, if these two issues could be addressed, a program of research could begin to genuinely examine the phenomena of awe and wonder as they are experienced in space flight.

The simulation for Experiment 1 was conceptualized as a truly immersive experience, where a range of emotions could be elicited at various stages of the simulation. To create awe and wonder experience while maintaining control of experimental conditions, designers addressed immersion, point-of-view, and practical venue concerns (Sherman & Craig 2002). The simulation designers wanted participants to engage in a willing suspension of disbelief, surrendering to the experience.

The design of Experiment 1, as we explained in Chapter 3, also tried to leverage the affective aspects of narrative, leaning on a framing narrative to support the sense of immersion (Figure 4.3). Research assistants explained that there would be a "launch" and participants who met the inclusion criteria (e.g. not color blind, right handed), were whisked into a mixed-reality space, made to resemble a space capsule, with portholes (served by LCD screens) that would provide digitally-generated views of the earth from space. The participants were fitted with physiological sensors (EEG, ECG, fNIR), completed computer-based surveys, and then "launched" (which involved light and sound changes). After launch

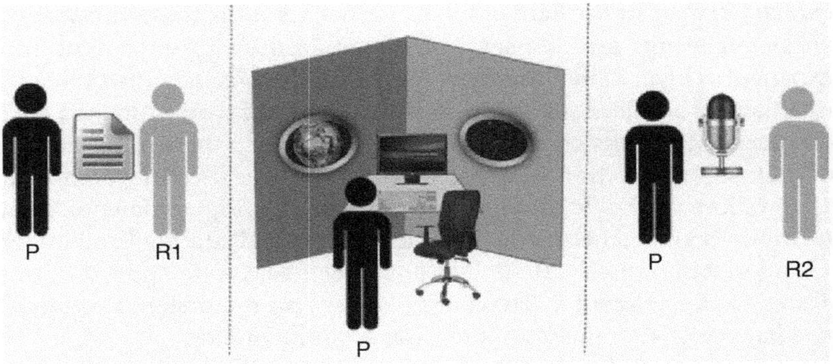

Figure 4.3 Experiment 1 was conducted in 3 phases. Left, the participant (P) completed informed consent and other eligibility paperwork with Researcher #1 (R1). Once eligibility was determined, P was moved to the spacecraft (middle) and prepared for "launch" by completing a battery of questionnaires and being fitted with EEG, fNIR, and ECG. After completing the simulation and all additional questionnaires, the participant met with another researcher (R2) for the phenomenological interview.

when the capsule was "floating in space," each porthole was opened consecutively for 12 minutes each showing two different space views, respectively (from among Earth view, Deep Space view, Earth view with IIS, or Deep Space with moon), with a short break in between to fill in questionnaires. After viewing the stimuli, participants were "radioed" by mission control and prepared for "re-entry." Once the simulation and additional questionnaires were completed, research assistants removed the sensors and escorted the participant to an area for the phenomenological interview. At that point in the experiment, the participant and the interviewer met for a recorded conversation where the interviewer would encourage the participant to describe her experience in the simulator.

Experiment 1 gave researchers a strong, positive indication in answer to question (1): Is it possible to elicit awe and wonder experiences in a simulated space environment? The psychological and neurophysiological measurements were collected and analyzed by human factors psychologists and neurophysiology experts. Phenomenologists analyzed the transcripts and recordings of the interviews, applying hermeneutic techniques. The phenomenological analysis was informed by the 34 consensus categories of experiences related to awe, wonder, curiosity, and humility. The results suggested that such experiences can be elicited in a simulation. The neurophysiological data showed a significant difference in theta levels, in both hemispheres, between participants who reported experiences of awe while viewing images of the earth and those who gave no indication of such experiences. In addition, beta activity was different for people viewing the Earth than when were viewing Deep Space. Furthermore, participants who scored higher on scales of religiosity (as measured by surveys) were more apt to report awe and wonder in conditions where the earth was seen from space. These relationships were new contributions and confirmed that, yes, experiences of awe and wonder could be elicited in the simulated environment.

The second question – How can we rigorously examine these experiences from the NP perspective? – remained less clearly answered. Were we rigorously examining these experiences from an NP perspective? Rigor should lend itself to replicability and predictability. While the data had shown some significant values, it fell short of what the researchers had hoped to discover. In addition, the experimental design could be criticized for not being fully neurophenomenological. After all, the human factors and physiological psychologists simply passed the participants on for a phenomenological interview after what could arguably be considered a straightforward laboratory experiment in the tradition of cognitive science.

Following Experiment 1, it became clear that methodological refinement could improve the statistical and inferential power of the study. The researchers conducted a post-experiment evaluation, a sort of post-flight debriefing, to determine what steps could be taken to improve the methods. This evaluation resulted in a range of suggestions for a subsequent experiment. Many of the suggestions were logistical refinements, applicable to this specific inquiry. For example, the evaluators suggested reducing the amount of time spent completing surveys to reduce incidents of boredom during the simulation. This is the sort of refinement that comes with experimental iterations, as human-subject researchers often face unanticipated responses in laboratory conditions. In this case, many psychology students found the lengthy surveys combined with the calm, quiet serenity of viewing the stars to be a recipe for sleepiness. That kind of refinement in experimentation is common in cognitive science research. However, the post-experimental evaluation also rendered lessons learned that spoke to the neurophenomenological approach more broadly (Bockelman et al. 2013). These lessons would apply to most NP studies, because the aim of NP is to infuse phenomenology into the experimental design and analysis while simultaneously assuring that the research is conducted in a manner that can be replicated or expanded by other scientists (Table 4.1).

Table 4.1 Lessons from the Experiment 1 and their implications for research

	Lesson description	Lesson application
Lesson 1	*A "shared mental model" must be intentionally developed and maintained across the key research team members*	1. Team training to prepare research assistants in phenomenology 2. Increased communication efforts to support interdisciplinary collaboration and coordination
Lesson 2	*The scientific study of experience needs to be positioned between the more holistic aspirations of enactive NP and legitimate strictures of scientific method that intentionally impoverish experiences to attain variable control.*	1. Higher control of variables through simulator modifications 2. Higher sample population 3. Emphasis on portability in simulation design 4. Development of new metric to compare phenomenological and psychological reports of experience
Lesson 3	*The impetus for phenomenological training is on the interviewer, not the participant, so that the interviewer may act to support the participant in precise experiential reporting.*	1. Implemented systemic training accountability for interviewers 2. Conceptualized the "training-tradeoff" to capture the shift of cognitive burden from participant to interviewer

The three step procedure for phenomenological research suggested by Varela (suspending beliefs or theories about experience, establishing intimacy with the domain of investigation, and seeking intersubjective validations) provided points for anchoring these lessons. There was no escaping that, when there was something to be improved upon, it was coupled to one or more of these principles. In this vein, the following lessons are presented as drawn from what we learned by doing Experiment 1, with examples for applying the lessons modeled in a similar follow-on study. The idea is to offer an opportunity to learn from our experiences, to show what we did the first time, and to indicate how the approaches could be improved to continue advancing a non-reductionist program of research into experience.

3.1 Lesson #1

A "shared mental model" must be intentionally developed and maintained across the key research team members.

This first lesson is clearly bound to the phenomenological challenge regarding intimacy with the domain of investigation. The concept of what team cognition scientists call a "shared mental model," or what phenomenologists might call a "shared understanding," however, takes this to another level, one that embraces the interactive components of an investigation as having equal importance with the domain of investigation. Getting this right is not just about the subject matter of inquiry, the experiences of awe and wonder, but about the people, tools, contexts, and intersubjective dynamisms as well.

The neurophenomenological perspective is an inherently interdisciplinary one. Like all interdisciplinary work, it requires the contributing disciplines to find ways to fit their approaches, not only with one another, but also to a shared objective. This is team science; it demands more than "passing the baton" among members of a group. Genuine collaboration and coordination in team endeavors require clearly defined roles and shared understanding of the goals.

This might seem close to common sense, but the actual dynamics of gathering a research team together and putting them all on the same page with respect to the experimental goal is often a practical problem in interdisciplinary research. All researchers come to the project with different assumptions that are directly tied to their own disciplines and to their research backgrounds.

Consider a common problem that has plagued good robotics design. Robots are meant to be embodied and pragmatic – physical devices designed to perform useful actions in the world. In designing the cognitive architecture for robots, however, scientists have often fallen back on their standard conceptions of cognition. The dominant idea with respect to building machines that can deal with high-level tasks is often still the representationalist stance of GOFAI (Good-Old Fashioned Artificial Intelligence – Haugeland 1985). This classic computational approach conceives of cognition in linear terms of sensing-thinking-acting – what has been called the "sandwich" model, where the "meat" in the sandwich is the thinking (cognition), and the relatively less important pieces of "bread" are the sensory input and motor output on either side (Hurley 1998). These kinds of cognitive control architectures that emphasize the "thinking" part are conceived as involving the building and updating of internal world models to be used for planning the next action. Sensory input and motor output are regarded as linear interfaces with the real world and are therefore considered as less challenging research problems. The consequence in robotic design is often that real-time responsiveness is lost.

> Properties of physical bodies were completely neglected in the strand of robotics that came out of the GOFAI tradition. Even in more recent and much more impressive examples that do successfully interact with the world in real time...a clear separation between body and brain is apparent. Indeed, the design philosophy was: "*Treat autonomous navigation as a software problem.*" (Dominey et al., in press).

The representationalist view of cognition shares with traditional systems engineering a reliance on top-down design and modularization. Processing is designed to be central and sequential; a collection of modules are linked to perform their function in a linear fashion (sensing-thinking-acting). On the one hand, the engineering disciplines favor this approach because it allows for clear definition of sub-tasks and division of labor among different units. The problem is that this applies not only to the design of the robot, but also to the design of the research project. When a research group goes to build a robot, each subgroup focuses on their own narrow task, creating a particular functional module that will be added to other modules that compose the system. Each module works perfectly well, doing precisely what it is designed to do. Unfortunately, the modules don't play well together. What happens is that no one has considered the relational aspects of how one module will couple with another to produce something that goes beyond a clunky machine-like behavior. In the more advanced cases, the project is meant to produce an autonomous robot expected to smoothly interact with humans. Humans, however, do not rely on the same kind of architecture. Instead, human behavior operates in dynamical attunement with the environment; it emerges on the basis of a dynamic coupling between organism and environment. As a result, even if the theory is updated to acknowledge a more enactive approach to robotic design, the actual practice of the researchers, working within the frame of their own particular paradigm, undermines the overall plan. If this point applies to scientists and engineers engaged in a clearly defined project, imagine how it applies to a project that attempts to integrate disciplines from the arts, humanities, and sciences.

The project needs to aim for a coordinated or collaborative, rather than a subordinated, hierarchical, or top-down form of management (Elias and Fiore 2012). Collaboration is a key feature of team cognition; it supports the emergence of a collective outcome that is greater than the sum of its parts. In team science, the complex interactions of individuals' unique talents and backgrounds (collaboration) combine with shared disciplinary egalitarianism (coordination) when the team, as a whole, centers on a shared goal. Keeping that goal in mind in the real nitty-gritty where artists and engineers are designing simulations to be used by psychologists and neuroscientists in experiments that generate data that, in part, require interpretation by phenomenologists and specialists in hermeneutics creates a problematic communicative situation.

The first lesson learned from the analysis of Experiment 1 was that neurophenomenological researchers must intentionally engage in the hard work of establishing and maintaining explicitly shared mental models (a shared understanding of all tasks) within a study. Otherwise, the results end up looking like a clunky robot. For example, we learned that the notion that one researcher or group of researchers simply completing a task and then passing the baton to the next researcher or group did not work very well. In Experiment 1, one part of the team went off to build the simulation, fully informed of criteria provided by the *Bildakt* group, and the experimental requirements stipulated by limitations of the neuroscientific technology. A large self-correction was required when the first part of the simulation design failed to incorporate some of the image criteria and the simulations appeared relatively flat. A second self-correction was needed when part of the simulation encouraged participants to move when the neuroscientific equipment required them to remain still. These seeming bumps in the road not only slowed our progress, but also suggested that we did not have a shared understanding of the project as a whole. This doesn't mean that philosophers must become neuroscientists (or vice versa) to engage in NP research. Rather, parts, roles, and responsibilities must be understood in the context of the full interdisciplinary goal of the given study.

Researchers applied this lesson in a subsequent experiment by intentionally engaging researchers in the philosophical and technical aspects of the project from the beginning of the experimental design and by working to maintain that communication to sustain a shared mental model throughout data analysis and interpretation. To reduce the "passed baton" aspect of the design, the follow up experimentation used a smaller number of researchers to collect the data (Bockelman et al. 2013). There were four graduate-level researchers, all with experience in human factors psychology and neurophysiological data collection. To assure that each of the data collectors understood the interdisciplinary mission to better understand first-person experience (as opposed to simply collecting cognitive science style correlations), the same researchers also engaged in phenomenological training, and specifically training focused on the phenomenological interview (see Lesson

#3, below). This also included ongoing discussions regarding the philosophical import of the study. Likewise, the phenomenologists on the team were brought into discussions to better understand some of the experimental design challenges inherent to immersive simulations so that there could be a shared understanding in regard to limitations of stimuli and their relationship to the clarity of the data (see Lesson #2, below). The principle is that a more integrated team will produce more integrated results.

The procedures for developing and maintaining a shared mental model cannot be prescribed in advance – they depend on the composition of the team and their specialties. There is no one-size-fits-all approach to designing and maintaining shared mental models (Cannon-Bowers et al. 1993). We did incorporate specific techniques in regard to Experiment 2 that helped to refine our methodology. For example, before designing Experiment 2, members of the team participated in concept mapping (Novak & Cañas 2006). Multiple members of the group discussed and challenged the relationships between all of the moving pieces. This helped the experimental designers better understand the intricacies of this specific study and it changed the way they could speak to particular issues throughout the experimental process. We also developed a practice of engaging in regular and ongoing training for the researchers so that the scientists who were well trained in traditional cognitive science methodologies could understand what phenomenology added to their toolset. We read and discussed basic tenets of phenomenology throughout our practice and training for interviewing (see Lesson #3, below). These efforts are time consuming, but they contributed to a productive continuity during experimental execution, which, in turn, contributed to higher confidence in data interpretation.

3.2 Lesson #2

> *The scientific study of experience needs to be positioned between the more holistic aspirations of enactive NP and legitimate strictures of scientific method that intentionally impoverish experiences to attain variable control.*

Although our neurophenomenological experiments have features of, and contributions from a range of disciplines (e.g. art history, hermeneutics, psychology, as well as neuroscience and phenomenology), its aim is to facilitate scientific knowledge. As is the case with all interdisciplinary scientific efforts, it is vulnerable to complex problems that emerge when there are lots of moving parts.

The results from Experiment 1 (see Chapter 3) offered something new to the study of awe and wonder, but some of the design decisions may have decreased the explanatory power of the results. Many possible variables (e.g. the time-sensitive launch count-down, the narrative detail, the mixed-reality components of the simulator, changes in setting and personnel between simulation and interview) introduced qualifications and complications in the data analysis. Since there were several variables that could account for participant experiences, researchers could not conclusively determine which of the manipulations generated the precise aspects of the experiences. As NP becomes more ambitious, and as more complex stimuli are used to address more complex human experiences, researchers confront a messier set of problems than were originally mapped out in the early Varela experiments (Lutz et al. 2002). The practical issue becomes one of maintaining clean experimental control as the conditions become more complex.

To some degree, this problem is tied to the first lesson. Members of the research team were indeed working from their own disciplinary perspectives and were not always cognizant of the impact that some of their small decisions had on the research agenda as a whole. The introduction of a framing narrative, for example, entered into the design to satisfy the simulation engineers who were concerned about making the simulation as realistic as possible. That intention was motivated by our overall concern for replicating the precise experiences of awe and wonder as they are tied to space travel. Yet, the use of a framing narrative and launch sequence clearly complicated anything we could say about what led to the replication. The second lesson, however, also involves a more basic issue, namely a tension between the strict, but very productive, limits that define scientific method (concerned especially with replication, reliability, and generalizability) and what might be called a philosophical holism that forms the background motivation for neurophenomenology. The issue is how to combine established methods in psychology and the cognitive sciences that provide rigorous and respected procedures for experimentation with the philosophical motivation for ecological validity found in neurophenomenology.

In part, this problem involves NP's insistence on addressing issues concerning first-person experience that are usually dismissed by cognitive psychology. That's the motivation of philosophical phenomenology as it pushes itself into the experimental setting. More than this, however, NP is the brainchild of the embodied, enactivist approach in cognitive science. Embodied cognition, and especially the phenomenologically-inspired enactivist approach to embodied cognition, contends that experience is not just a matter of neuronal processing in the brain, but that it involves bodily (motor, affective, autonomic, and peripheral) processes that are constrained by the organism's coupling to its physical, social, and cultural environments and by the organism's (the subject's) previous history. Such an approach is not only non-reductionistic, but is also concerned with the dynamical connections among all of these factors, and with ecological veracity. All of this brings it directly into conflict with the important quest for methodological control in science.

Certain elements in Experiment 1 were already compromises between the enactive view and the scientific requirements for controlled experiments. For example, enactivism emphasizes the role of sensory-motor contingencies (e.g., O'Regan and Noë 2001). Moving around in the environment affords more nuanced perception and even the potential for movement modulates the agent's perception. The VSL was designed, however, to minimize movement and to focus the subject's attention on the virtual space scenes. While this in some very real sense structured the kind of coupling that took place between agent and environment, it did so in a way that limited any effect of bodily movement. This limitation can be justificd, however, by the fact that, in some circumstances, astronauts are also limited in their bodily movement when, for example, they are strapped into a workstation on the IIS. Beyond sensory-motor contingencies, however, other bodily processes influence perception (for an enactivist view on this, see Gallagher and Bower 2014), and we had imperfect or incomplete measures in place for those aspects. ECG did give us good information on heart rate, etc., but any attempt to add further measures would add to the already significant complexity of data analysis. So, while we had some good measures of the bodily processes that enactivism holds to be important for shaping experience, they were still more limited than the enactivist might want.

More generally, these trade-offs or compromises reflect the tension between doing the rigorous science of human experience and the embodied/enactivist contention that cognition and experience are influenced (and often constituted) by extra-neural factors of body, environment, and prior history (embedded in social and cultural contexts).

Embodied/enactivist approaches to cognition present a challenge for science. By focusing not just on the brain, not just on the environment, not just on behavior, but on the rich dynamics of brain-body-environment, enactivists offer a holistic conception of cognition. To put it succinctly, however, it is difficult to operationalize holism. Neither experimental control nor the division of labor in science allows for all factors to be taken into consideration at once (Gallagher, in press).

If neurophenomenology is a scientific research program, the enactivism that stands behind it is more like a philosophy of mind. On the one hand, enactivism makes empirical claims, for example, about the work of sensorimotor contingencies, and in this sense it resembles a research program that can suggest new experiments and new ways of interpreting data. On the other hand, its emphasis on holism presents problems for empirical investigations. One doesn't get far in experimental science without controlling for variables.

Enactivism takes seriously the results of science, and its claims remain consistent with them. However, its role may be better understood as offering critical distance and keeping science honest – something that is important for the larger project of science. To the extent that it can offer practical suggestions, in some cases it may make doing science more difficult. Enactivism does not endorse the mechanistic definition of the mind, or of nature more generally, presupposed by science, but contends that the mind and nature cannot be understood apart from the cognitive capacity that we have to investigate it. In the context of a philosophy of mind and nature that is meant to offer an encompassing view, holism should be viewed as a strength rather than a practical complication (see Gallagher, in press) But at the same time, when one is engaging in science, it is a practical complication.

As we can see in Experiment 1, enactivism, as it figures into neurophenomenology, may still motivate experimental science in very specific ways. Even if, in some cases, it is difficult to apply a holistic view to a given question, there may not be any special complication in designing experiments that can test enactivist ideas. The lesson is that, on the one hand, we should not, in every case, include absolutely everything that may be relevant to dealing with a particular concrete question. However, on the other hand, in the end, it may be easier to include some difficult but crucial factor than to ignore it. Although, in this and other cases, much will depend on circumstances like the availability of the right lab technology, the complicated whole may sometimes lead to simpler explanations. Still, it is no easy task to design rich, dynamic, complex, and life-like conditions that are also highly controlled.

Having said all of this, it's still necessary to integrate the phenomenological and the enactive with the third-person approaches that cognitive science has developed. NP, just like more standard cognitive science, must consider questions of generalizability, verification, and replicability. For example, confident replication would be difficult with respect to Experiment 1. This problem, however, can be overcome by making better use of simulation technologies. Simulations vary from very low fidelity (e.g. a board game like *Life* or *Monopoly*, which reflect lifestyle or market decisions) to high fidelity immersive experiences (e.g. immersive amusement park rides, where the visitors interact with intelligent agents, can smell, feel, and see various layers of entirely synthetic elements mixed with real objects and actors). Depending on the simulation design, research institutions can share systems, relocating the equipment or transferring the software, so that the stimuli can be precisely controlled and presented to various populations. This not only supports replication, but also generalizability.

Experiment 1 used a mixed-reality simulation successfully, and it helped create an immersive experience. However, the degree of immersion may have been confounded by various changes of researchers (see Figure 4.3) as well as multiple elements in the environment (e.g. the participant's tactile interaction with the computer, the simulated radio communication from the base, the visual presentations from two "windows") without fully exploring the role each element played. Experiment 2 streamlined the modalities (there was only visual simulation), and the interactions (there was only one researcher with whom each participant interacted throughout the experiment), and conducted all parts of the experiment in one location (Figure 4.4).

In Experiment 2, we also used a portable simulation environment, supporting collaboration and replication. Simple, projection-based simulation environments can be brought to sample populations and offer a cost-effective form of highly-controlled presentation. Digital artists, working closely with other members of the team, including our NASA astronaut consultant, Jeffrey Williams, were able to design the stimuli with high degrees of control and visual fidelity. The end result was a highly immersive (some participants reported feeling motion and/or hearing engine sounds although there was no kinetic or audio presentation) and highly replicable test bed. The conditions were better controlled for precise run time, differing only in visual perspective (see Chapter 5), and the variances between experiences could be confidently attributed to the stimuli rather than changes in location, narration, or personnel.

3.3 Lesson #3

For participants who are not trained in phenomenology, the interviewer supports the participant's efforts for mindful acknowledgment and articulation of experience.

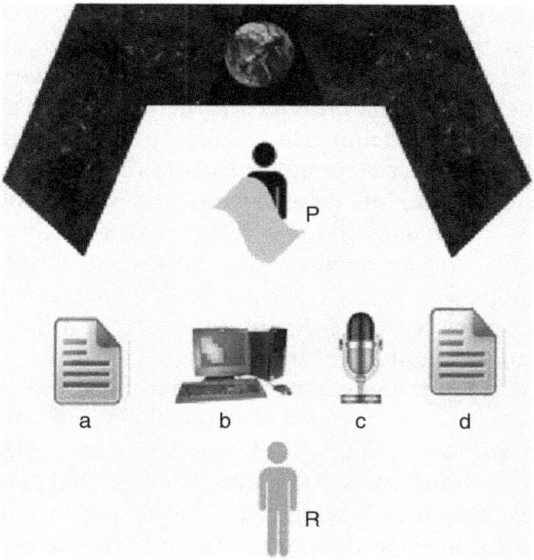

Figure 4.4 In Experiment 2, participants (P) interacted with one researcher (R) in one location with a visual-only immersive simulation: a) informed consent and eligibility requirements were given; b) R applied physiological measures and monitored from an out-of-view position behind the seated P during the stimulus presentation; c) physiological equipment was removed and R & P sat together in the simulation space during interview; d) P completed psychometrics.

In NP studies where subjects are trained in phenomenological methods (e.g. Lutz et al. 2002), the phenomenological interview (used, for example, to develop the phenomenological clusters) is important, but its role is somewhat offset by the careful and systematic training of the participants to engage in phenomenologically mindful practices. The participants are taught *how* to precisely and effectively reflect on and articulate their experiences.

Such training takes time. Unfortunately, for researchers (especially at universities where the studies are time constrained and participants may be available for no more than a few hours total), time is not a limitless resource. Furthermore, not all participants are equally as good at reflecting on and reporting their experiences, even if all are equally trained. One has to worry about the possibility that reflectively accessing experiences can transform those experiences into something different from what they originally were. Phenomenological method attempts to minimize such transformations, but again not all who are equally trained are equally good at avoiding such transformations.

Experiment 1 provided a clear lesson in ways to improve the phenomenological interview so that untrained participants could gain access to their experiences and find means to articulate them. Shifting some of the burden of reflective awareness and articulation from the participant to the experimenter during the interview can address the problem of time constraint and can reduce the possibility of reflectively transforming experience. This shifting requires what Bockelman (2013) referred to as the "training trade-off"; the perspectival differences between interviewer and interviewee can be significantly reduced *if* the methodological training and accountability are sufficient.

The importance of the phenomenological interview in making available first-person data moves in the opposite direction to the constraints imposed in response to Lesson #2. Where that lesson resulted in tightening up objective measures and controls, addressing Lesson #3 gives us a way of dealing with the subjectivity that we want to study, without reducing it to third-person data, and without having to become subjective in our approach. This is where the neurophenomenological method pushes most strongly against reductionism by maintaining the subject and its phenomenological analysis. The phenomenological interview allows the interviewer to participate in the reflection on and articulation of experience, and if this stands in contrast to the third-person objective methods otherwise employed, it also opens doors for new data collection and analysis techniques.

In our experiments, we used techniques similar to those employed in the NP studies of epilepsy, but unlike the work with patients who could be visited on numerous occasions, we had one opportunity to meet with each participant to collect all of the necessary information. One of the observations from Experiment 1 was that participants are not simply "good" or "bad" at examining their own experiences. Rather, they fall along a continuum. Upon review of the transcripts, it became apparent that some questions asked by the interviewer, and some of the interviewer's responses to participant articulations influenced subsequent participant statements. Straightforward techniques like affirmations from the interviewer (e.g. nodding, saying "Alright" or "Uh huh") or reflective language can lead participants to describe their experiences in various degrees of detail. Some questions led to more descriptions than other questions.

These observations informed the "training trade-off" hypothesis. The research team came to realize that the interviewers must carry more responsibility for extracting non-judgmental experiential accounts when the participant is untrained in practices of self-awareness. However, the degree of participant phenomenological acumen (i.e., their ability for self-reflection and mindful non-judgment reports) cannot be controlled in most randomized studies. Consequently, the interviewers should be trained to anticipate the lowest level of participant performance in this area and be prepared to support reflection and articulation. The training trade-off compensates in that the interviewer's skill makes up for deficiencies that participants may have in this area; "If the interviewer is working with Buddhist monks, she may not need to receive a great deal of training and may be able to simply tell the participant the focus of the study. Conversely, if the same interviewer is working with undergrads at any given university in the West, she may need to pull from a collection of tools and techniques to give the participant the capacity to access the thoughts and feelings experienced" (Bockelman et al. 2013).

The training trade-off is primarily about method, but the theoretical contribution of the hypothesis deserves attention with respect to the larger antireductionist stance. Performance in the interview hinges on the level of established dynamism between the interviewer and the participant in a manner that stands in opposition to the notion that legitimate cognitive processes are entirely brain-bound. In the NP interview, there is a cognitive off-loading of participant to interviewer, as the interviewer elicits the experiential account. The interviewer does some of the cognitive work for a phenomenologically untrained participant that otherwise a trained participant could do independently. The interviewer focuses on the participant's lived experience in a manner that the participant may not be able to do on his own. The interviewer uses tools of phenomenological interview, in a manner that deviates from traditional cognitive psychology.

Critics of the neurophenomenological technique argue that science requires third-person objective data; they reject the value of first-person experience and advocate a detached objectivity on the part of researchers (Dennett 1991). Dennett worries that (i) first-person reports can be biased or inaccurate; (ii) the process of generating first-person reports about an experience can modify that experience; and (iii) there is an "explanatory gap" in our understanding of how to relate first-person, phenomenological data to third-person, biobehavioural data. Lutz and Thompson (2003) respond to and reject such concerns about a first-person neuroscience, addressing issues of bias in first-person reports, the transmutation of experience in its recounting, and the relationship between phenomenological data and biological/behavioral data. The recommendations for the phenomenological interview are grounded in their arguments.

If properly executed, the phenomenological interview gives the researcher a way to leverage the interactive and shared components of communication to preserve first-person experience. The phenomenological interviewer has no pretense of objectivity and is not a mere observer. Rather, the interviewer assists and supports the participant so that experiences may be reflected on without judgment and expressed for data capture. For example, the interviewer:

1. Keeps the participant on target, so as to avoid digression (e.g. "You mentioned feeling excited when you saw the ocean, could you tell me more about that?")
2. Encourages more detailed explications that explore articulations that may be unclear (e.g. "You said that the stars were "cool." What do you mean by "cool"?)
3. Encourages descriptions that reference the whole body (e.g. "You said that you felt "queasy" when you were moving away from the earth. Can you describe what that feels like? Where are these sensations in your body?")
4. Leverages the environment to promote recall (e.g. "Feel free to gesture toward the screens at any time to indicate where things were on the screen at the time you felt this way.")
5. Consistently uses second person pronouns to direct participant's inward focus.

If the interviewer is sufficiently trained, the participant need not be trained at all. The interviewer is doing a share of the work required for reflection and articulation. She participates interactively in the explication of the experiential data, but with the intentional, systematic use of cues and articulations that resist biasing the participant. The interviewer does not engage in mindreading (understood in the terms of social cognition), does not ascribe mental states to the participant, and has no need to access personal mental models of experience in order to facilitate the participants' accounts.

Traditional cognitive psychology establishes a framework in which the participant must make his reports conform to pre-specified requirements of the experiment. The questions are uniform and control means consistent conforming to a template so that everyone is treated identically. The assumption is that the data collected is more reliable because it has avoided bias. Actually it has only adopted the bias of the experimenters – which may be productive, but which may also be constraining. The phenomenological approach assumes that subjective bias can be mitigated using focused techniques by highly trained interviewers, and that the variance in things like, length of interview, number of questions, and topics discussed reflect the variances in the experiencers and their experiences. The resulting experiential record is more complex and complete than that which would be collected by a template driven approach of traditional cognitive psychology.

Preparation for Experiment 2 included a more extensive training regimen in which the data collection researchers engaged in the following: theoretical discussions regarding phenomenology and its tools for exploring experience, analysis of the interviews from Experiment 1 to examine strengths and weaknesses, instruction on second-person (interviewer-assisted) reflection techniques, and interview rehearsals. By the time data collection began, the interviewers were familiar with barriers to maintaining reflective focus and describing experience. The interviews involve the sort of reflection that is not merely inward-directed thought, as if the experiences were not situated in a rich and meaningful world. Reflection is also concerned with the intentionality of experience – not only the "what it is like," but the what it is that one is experiencing. References to landmarks in the simulations were important for latter correlation with the simulation timeline.

This chapter has endeavored to advance the discussion about how to *perform* an enactive, neurophenomenological science, one that incorporates third-person objective and first-person subjective data in ways that retain the statistical power of established methods while embracing the inherent value of first-person reports of experience. These considerations were tested in a second experiment, and we turn to that in the next chapter.

5
Redesigning Plato's Cave: Experiment 2

Experiment 1 had confirmed a number of important points for our project. First, we were able to elicit experiences of awe and wonder that replicated (or came close to) those articulated by the astronauts. Second, the specific combination of methods in a neurophenomenological framework showed potential as a way to explore those experiences. As we detailed in Chapter 4, however, the first experiment presented some methodological challenges along the way. In this chapter, we describe how we responded to the lessons learned from Experiment 1. We describe our second experiment and develop a style for data integration that considers physiology, psychology, and phenomenology as a combination of approaches that can offer a precise triangulation on the experiences in question.

1 Re-designing space on earth

Although there were a number of experimental features that we wanted to change for the next round of experimentation, one thing that remained consistent was our desire to continue leveraging simulation technologies for the advancement of the project. The VSL, in its contextually-rich immersion, had transported some participants to a place where they could experience awe and wonder. Their accounts during the phenomenological interviews suggested that the mixed sensory modalities involved in that simulation were all playing some role. Some participants credited the visuals as the key catalysts for such experiences, while others pointed to a myriad of features as generating the sensations they experienced during the experiment. In some respects, this was good – just as in the real world, different people appeared to respond to different things.

However, this wasn't "the real world." This was a laboratory experiment; and, as such, we wanted to understand the relationships between specific aspects of viewing the earth from space and the spiritual and

Plate 1 Interior of the VSL

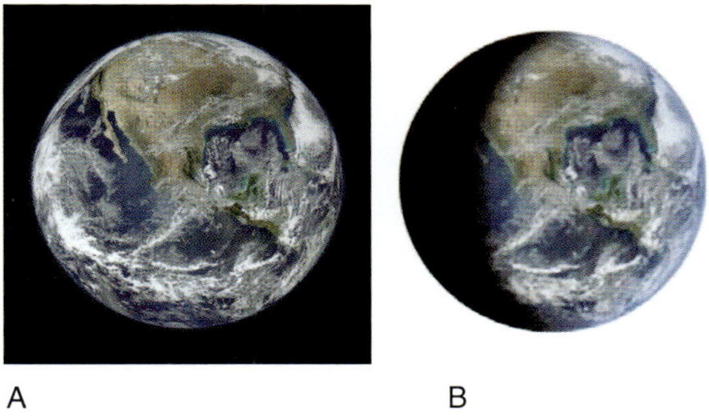

Plate 2 (A) Blue Marble 2012 – NASA image; (B) Blue Marble modified

Plate 3 The FOC-condition began near the earth, over a view of the participant's university

Plate 4 The FOC-condition pulled away from the earth, while revolving

Plate 5 Final vantage in the FOC-condition

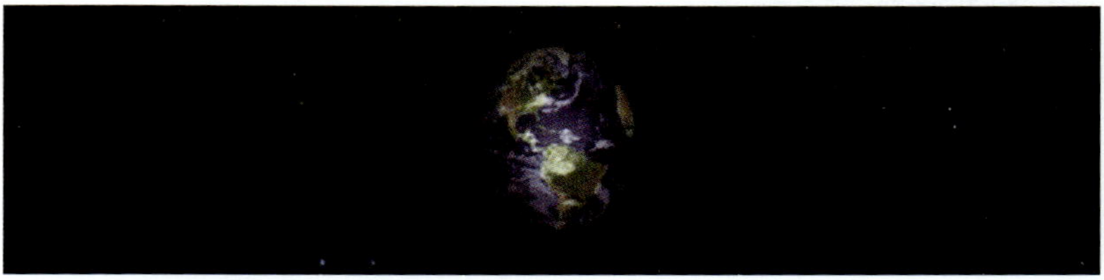

Plate 6 The final vantage in GLO-condition

Plate 7 Giorgione's *La Tempesta*

Plate 8 *Destination Moon* (1950)

Plate 9 The 'Blue Marble' photograph, taken on December 7, 1972, by the crew of the Apollo 17 spacecraft at a distance of about 29,000 kilometres. It shows Africa, Antarctica, and the Arabian Peninsula. Public domain image. http://www.nasa.gov/images/content/115334main_image_feature_329_ys_full.jpg

Plate 10 The Earth-Moon System (http://solarsystem.nasa.gov/multimedia/gallery/Earth_Moon_br.jpg)

affective responses involving awe and wonder. Consequently, we redesigned the simulation test bed to focus on the visual components alone.

The visual conditions in the VSL were approximately 12 minutes each, with each participant viewing two of four vantages. While some participants did report experiences of awe and wonder involving spiritual or aesthetic aspects, many also reported that they had grown bored during the experiment. For the second experiment, we wanted to shorten the simulation time so that any physiological signs of relaxation could be credited to the stimulus alone, and not to possible boredom or fatigue.

In the second experiment, then, we moved from the mixed-reality to a purely virtual-reality simulation. The test bed was located in a dedicated room with a cave-style 120° field of view. Digital artists generated the simulation, using a mixture of satellite imagery and 3D graphical artistry. For the philosophers involved in this study, this was an updated and high-tech version of Plato's cave.

The contents of the simulations were streamlined as well to restrict the sensory modality to only visual and to control for context. We wanted to understand more about the experience of seeing the earth from space, but it is clear that never happens without some context. There are years of training and anticipation before a person has the opportunity for space travel. Then, the astronaut has a natural unfolding context that situates the experience of seeing the earth from space. Context is a combination of what the subject brings to the situation and what the situation has to offer the subject. Each subject carries his or her own past experiences and expectations into the simulation; but what the simulation has to offer – what we can call the "presented context" – is something we could control. For our participants, the presented context became the critical control. We used two simulated space-flight conditions, each seven minutes, and each participant was exposed to only one of those conditions.

The first condition started the simulation hovering over the University of Central Florida campus where our participants attended classes, giving us confidence that it was the one location on earth we could assume was familiar to all participants (Figure 5.1). We note that the aerial view of the UCF campus is somewhat iconic (arranged in a circular pattern), and is familiar from campus maps provided to students and visitors to campus. From this focal (FOC) view, the participant's perspective was pulled farther from the ground, until the image looked similar to the blue marble discussed in Chapter 3 (Figure 3.1).

The other condition removed the context of seeing the earth pull away from a familiar location. It began on the shadowed side of earth, so that the first view was of the crescent horizon (Figure 5.2). This global

(GLO) condition placed the participant in space, now with no visual or auditory cues leading up to that perspective.

These conditions were essentially composed from one simulation with a total run time of nine minutes. FOC-participants viewed only the first seven minutes, which allowed them to witness a movement away from the earth, whereas the GLO-participants watched a more distal scene by viewing the last seven minutes. Because of distortions in human perception of distance, the final impression of the earth was similar in both conditions (Figures 5.3 and 5.4), although the speed of movement away from the earth was held constant.

Figure 5.1 The FOC-condition began near the earth, over a view of the participant's university

Figure 5.2 The FOC-condition pulled away from the earth, while revolving. This figure shows the point in the simulation where the vantage has pulled back enough to see horizon. This is also the early imagery in the GLO-condition, as that condition started at a later point in the simulation timeline.

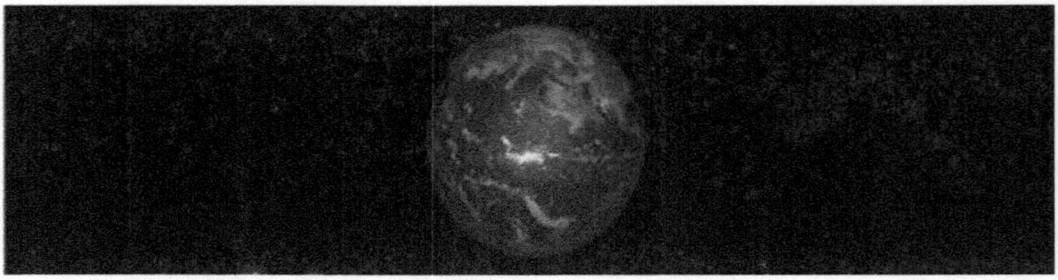

Figure 5.3 Final vantage in the FOC-condition.

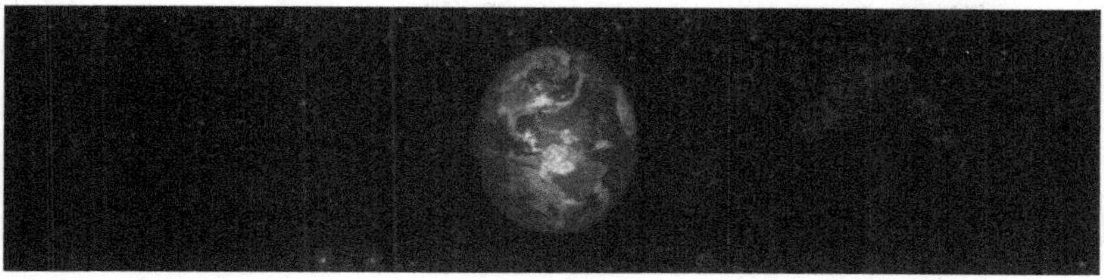

Figure 5.4 The final vantage in GLO-condition.

At completion, these synthetic video sequences were projected on the 7 ft. wrapping screens, in the dedicated test bed. We used the same trajectories and animation algorithms to create a control condition. During the control condition, participants viewed a geometric shape moving distally with similar levels of speed, light contrast, and direction as used in the experimental visuals. With this simulation set, we conducted the second experiment in the exploration of awe and wonder.

2 Methods

In a 1×1 experimental design (groups assigned between either a focal or a global view of Earth) with repeated measures on the first variable, each participant received a control and counterbalanced condition. That presentation order was consistent for the optimization of recall and the control of effect size from the magnitude of the viewing area. Groups were assigned randomly prior to participant arrival. We conducted experimentation entirely within a dedicated simulation test bed space. The participants sat in a low-profile (gaming-style) chair during the visual simulation, so that the natural vertical and horizontal range of view surrounded the periphery within the panoramic projection.

2.1 Participants

Seventy-four participants were recruited from the University of Central Florida psychology SONA system, ages ranging from 18 to 32. As in the first experiment, participants were screened on the following criteria: age, right-handedness, color blindness, seizures, tobacco or caffeine intake, alcohol consumption, and ingestion of antipsychotics or antidepressants. Total participation time was approximately 2.5 hours. Participants were all fluent English speakers with normal or corrected to normal vision.

2.2 Dependent variables

Measurements were collected through neurophysiological, psychological (i.e. surveys and questionnaires), and phenomenological sources. Dependent variables were measured along three disciplinary lines: physiology, psychology, and phenomenology. A description of the interdisciplinary tools follows.

Neurophysiology

Neurophysiological measurements used multiple sources of input: electroencephalography (EEG), electrocardiography (ECG), and functional near-infrared (fNIR). These tools provide high degrees of temporal sensitivity to change with minimal interference with the first-person experiences during stimulus presentation. The B-Alert X10 wireless EEG collected data from brain activity across nine channels with sensors placed bihemispherically in anterior, central, and posterior brain areas. Specifically, alpha, beta, and theta waves were recorded for EEG. The left and right hemisphere oxygenation was recorded using the Somantec INVOS oximeter, an fNIR device.

Psychology

A variety of questionnaires were administered. At the beginning of the session, participants completed the Ishihara Color Blindness Test (Ishihara 2010) to ensure typical color vision.

Prior to stimulus presentation, participants completed a computerized series of questionnaires. The questionnaires issued prior to the experimental conditions were selected to avoid priming and focused on personality traits pertinent to understanding awe and wonder. The Multiple Stimulus Types Ambiguity Tolerance scale (MSTAT) (McClain 2009) is a 22 item measure that determines an individual's tolerance for ambiguity. The Tellegen Absorption Scale (TAS) (Tellegen and Atkinson 1974) is a 34-item instrument that measures participants' openness to absorbing self-altering experiences in seven scales: Responsiveness to Engaging Stimuli, Synesthesia, Enhanced Cognition, Oblivious/Dissociative Involvement, Vivid Reminiscence, and Enhanced Awareness.

After the simulation and phenomenological interview, several other questionnaires were administered. We designed an experiment-specific questionnaire – the Experiment-Specific Survey of Experience (ESSE) (see Appendix). This questionnaire was developed to provide additional quantitative support for the pertinent constructs of awe and wonder, the consensus categories. The ESSE is a computer administered questionnaire that explicitly asked participants the degree to which they self-identify as a "spiritual person," "logical person," and/or "religious person." The ESSE also provided participants with formal definitions for awe, and wonder and related concepts (as described in Chapter 2). They were asked to what extent they agreed with a statement such as, "While viewing the presentation today, I experienced AWE." They indicated the degree to which they experienced awe, wonder, and other related constructs like humility and curiosity on a 100-point sliding Likert-scale. If participants indicated an experience greater than 10 (on the 100 scale), then they were issued an automatic follow-up regarding the time in the simulation when they experienced that category. For example, the prompt would read, "I experienced AWE the most when viewing: a) close images of the Earth (toward the beginning of the video); b) distant images of the Earth (toward the end of the video); c) the image of the geometric shape."

Finally, as in the first experiment, participants completed the Brief Multidimensional Measure of Religiousness/Spirituality (BMMRS) (Idler et al. 2003; Masters 2013; Masters et al. 2009).

Phenomenology

A post-simulation phenomenological interview (see Chapter 4) was conducted to collect first-person reports of participant experiences during the simulation. Immediately following the simulation, a research assistant interviewed the participant to collect the first-person experiential data. To support continuity, research assistants responsible for conducting the neurophysiological and psychological aspects of the experiment also conducted the interview. In order to conduct the interviews, they were trained to focus on descriptive terms and to seek clarity of those terms. Interviewers led interviewees to avoid judgments and self-analysis, as the desired report did not concern their opinions of their experiences, but rather focused on the descriptions of the experience. The interviewer thus helped the participant accomplish the suspension, redirection, and receptive openness prioritized in phenomenological methods. These methods allowed the participants to describe all of the experiences in the simulation in detail and provided first-person qualitative data.

2.3 Procedure

The inclusion criteria were provided to potential participants within the university's online recruiting system before a person registered for participation. Upon arrival, participants were confirmed for meeting the inclusion criteria and read the consent form. Informed Consent was provided to the participants before neurophysiological sensors were applied. A research assistant then equipped the participants with the neurophysiological sensors. The EEG cap with ECG electrodes was applied. Additionally, the research assistant fitted participants with fNIR sensors. This was the same suite of physiological sensors used in the first experiment. The next step was also one conducted in the first experiment; participants completed a five-minute resting baseline for the neurophysiological measures. This was required to calculate change scores for the neurophysiological measures during the scenarios. The baseline was conducted with eyes open, and participants were instructed to maintain a relaxed focus forward while no lights were on in the test bed area.

Participants completed the MSTAT, TAS and demographic questionnaires prior to the simulation. These psychometrics were chosen because the initial round of experimentation suggested that they would provide further insight, while the surveys that were no longer relevant were dropped from the experiment altogether or moved to the very end of the data collection, after the interview.

The participants observed two visual simulations. First, all participants completed a familiarization period to become acclimated to the simulation space by observing the control condition. This was to determine that responses were truly a result of the space simulation and not simply a response to the large-scale projection itself. Then, an experimental space (FOC or GLO) condition was presented, counterbalanced from one of the two manipulations of imagery of Earth described in the section above.

Upon completion of the simulation, the neurophysiological sensors were removed and a phenomenological interview was conducted to ask the participants about their experience during the simulation. The interview was audio recorded and saved for later transcription. The study was completed with participants answering the ESSE and BMMRS.

3 Results

The resulting data was analyzed through methods from each key contributing domain: psychology, neuroscience, and phenomenology. The goal was to use the tools of these three main lenses to re-focus the results from a distinctly non-reductionist perspective. Accordingly, this results section is organized by sequentially connecting each disciplinary perspective to another: first the psychological surveys are connected to the neurophysiological findings. Then, the neurophysiological results are correlated with the phenomenological findings. Finally, the phenomenological and psychological results are presented in light of one another. The discussion section will aim to synthesize data across these connections and explore implications from each of these perspectives.

3.1 Psychological and neurophysiological correlations

In this section, we provide an overview of the experimental results regarding neurophysiological activity during the observation of the space simulation. First, analyses were conducted to examine the effect of each condition on neurophysiological behaviors over simulation time. This addresses the experimental manipulations. Then, the neurophysiology is compared to self-reports from the psychological metrics. These analyses address the questions of the nature and structure of experience, as posed by the neurophenomenological method, by using psychological reports to assist in the interpretation of the neurophysiological findings.

Condition by minute for hemisphere

We began our neurophysiological analysis by looking at each minute of simulation time for the two conditions. A 2x7 (condition: FOC and GLO by minute: 1–7 min of simulation viewing) mixed ANOVA with repeated measures on the last factor was conducted for each EEG hemisphere by frequency (alpha, beta, and theta) and for each fNIR hemisphere RO^2 to identify any physiological difference between conditions and processing requirements for the duration of the simulation. At that first level of analysis, there were no significant results between the groups across any of the measures.

We continued the analysis by looking at differences between subjects, looking at alpha, theta, and beta wavelengths as captured in the EEG. The following results are those that reached statistical significance in that round of analysis.

The left hemisphere alpha differences between the FOC and GLO conditions were significant during the second minute F (1, 67) = 4.423, p = .039; FOC (M= –8006.92) < GLO (M= –4997.41). Significant differences were also recorded during the seventh minute: F (1, 66) = 4.040, p = .049; FOC (M= –6458.06) < GLO (M= –3731.10). (Figure 5.5).

During the second, third, and seventh minutes, there was a significant effect of condition presentation on left hemisphere beta. Left hemisphere beta during minute two was significantly different between conditions, F (1, 67) = 18.639, p .001; FOC (M= –1342.86) < GLO

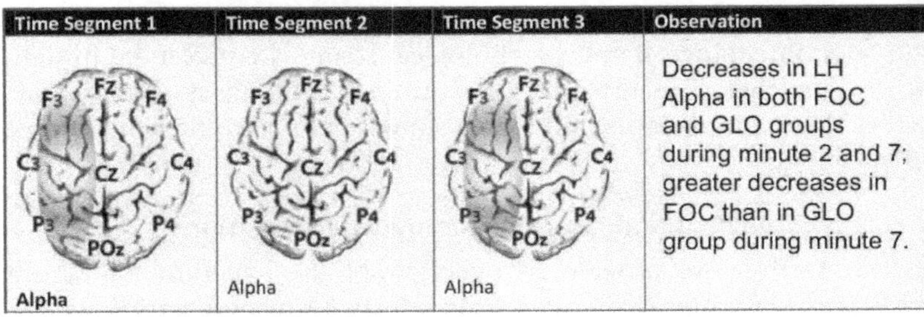

Figure 5.5 Decreases in LH Alpha

(M=388.10). A difference was found in this region during the third minute, F (1, 67) = 14.238, p = .035; FOC (M = –755.83) < GLO (M = –133.070) and seventh minute F (1, 67) = 6.368, p = .014; FOC (M = –914.825) < GLO (M = –184.498). No significant differences were found for minutes one, four, five, and six.

Significant differences between conditions by minute for the right hemisphere beta were found for minutes two, and three. In minute two, the difference was significant F (1, 67) = 17.245; p < .001; FOC (M = –1128.564) < GLO (M = 623.349) and minute three F (1, 67) = 5.647; p = .020; FOC (M = –609.296) < GLO (M =103.237).

Condition by psychological self-reports

Between groups, (FOC and GLO conditions) ANOVAs were run to determine if the visuals had an impact on experiences of awe and wonder. This was an important analysis to run because modifications in the methodology from Experiment 1 suggested that Earth views were more engaging and elicited greater affective responses. Therefore, this analysis helps determine the role of the image of Earth itself in eliciting these responses. No significant effects were found for the ESSE metrics on self-identification as "spiritual person," "logical person," "religious person," or "reflective person." The experiences of awe and wonder did not differ significantly between FOC and GLO groups, and there was no significant difference reported for feelings of familiarity.

Hemispheric behavior by psychological reports

Given that there were no significant differences between the FOC and GLO groups in their responses to the ESSE – that is, the participants were just as likely to self-report awe and wonder regardless of which condition they received, the following analyses were collapsed across both conditions to understand the relation between time, physiological response, and reported experience. This extends what we learned from Experiment 1, that amount of time in simulation is a factor in having experiences of awe and wonder.

Correlations were run between subjects who self-identified as spiritual, religious, reflective, and/or logical to check for construct independence to better understand the differences, or lack of differences between them. These correlations also become important later in our discussion of the phenomenological interviews.

Participants who self-identified as "spiritual" also tended to identify themselves as "religious" at a significant level ($r = .764, p < .001$). They also significantly indicated experiencing wonder ($r = .253, p = .037$), but there was no significant correlation with the awe, curiosity, or humility. Self-identification as "spiritual" also correlated with difference from baseline in theta activity in the left hemisphere during the second minute ($r = .259$, $p = .033$) and the sixth minute ($r = .264, p = .029$). Theta differences from baseline were also correlated in the right hemisphere for self-identified "spiritual" participants, with significant correlations recorded for the second minute ($r = .334, p = .005$), fifth minute ($r = .282, p = .020$), sixth minute ($r = .267, p = .028$), and the seventh ($r = .291, p = .016$).

The self-identification of "religious" correlated with awe ($r = .290$, $p = .016$), but not wonder, curiosity, or humility. It also correlated significantly with theta changes in the left hemisphere during the fourth minute ($r = .242, p = .047$), fifth minute ($r = .281, p = .020$), sixth minute ($r = .320$, $p = .008$), and seventh minute ($r = .257, p = .034$). In the right hemisphere, there were significant correlations with alpha difference from baseline during the first minute, ($r = .246, p = .043$) and second minute ($r = .252, p = .038$). There were also significant correlations between self-identification as "religious" and right hemisphere theta changes during the second minute ($r = .298, p = .013$), fifth minute ($r = .282, p = .020$), sixth minute ($r = .277, p = .022$) and seventh minute ($r = .269, p = .027$).

Like "religious," self-identification as "logical" correlated significantly with the survey-reported experience of awe ($r = .267, p = .028$). "Logical" also had a significant correlation with reported feelings of familiarity ($r = -.328, p = .006$), whereas "spiritual" and "religious" did not. It also had a negative correlation with right hemisphere theta ($r = -.249, p = .040$).

Correlations for visual processes

An analysis was conducted to capture aspects of the relationship between visual processing and first-person reports in the ESSE. As the simulation was exclusively visual, we analyzed areas involved in the processing of visual information, specifically the posterior parietal and anterior occipital lobes as collected from the P3, P4 and POz EEG sensors.

Participant self-identification as "spiritual" in the ESSE correlated with difference from baseline changes in alpha in these areas during the first minute ($r = .244, p = .045$), second minute ($r = .280, p = .021$), sixth minute ($r = .267, p = .028$) and seventh minute ($r = .330, p = .006$). Changes from baseline in theta also significantly correlated with "spiritual" self-identification for the second minute ($r = .319, p = .008$), fifth minute ($r = .315, p = .009$), sixth minute ($r = .267, p = .028$), and seventh minute ($r = .282, p = .020$). There were no significant correlations between "spiritual" self-identification and beta signals in these areas.

Participant self-identification as "religious" in the ESSE correlated with difference from baseline alpha in the posterior parietal and anterior occipital lobes during the first minute ($r = .245, p = .044$) and second minute ($r = .249, p = .040$). Beta changes correlated with this category during the seventh minute ($r = .277, p = .022$). Significant correlations also were found between those who self-identified as "religious" and those who did not self-identify as "religious" for theta difference from baseline in these areas during the fifth minute ($r = .291, p = .016$) and seventh minute ($r = .270, p = .026$).

Self-identification as "logical" in the ESSE negatively correlated with difference from baseline alpha for these areas during the seventh minute ($r = -.241, p = .048$). There was also a negative correlation with theta change in the last two minutes: sixth minute ($r = -.262, p = .031$) and seventh minute ($r = -.288, p = .017$). There was no significance found for a relationship with alpha in the posterior parietal and anterior occipital lobes and "logical" self-identification.

3.2 Neurophysiological and phenomenological results

The following results integrate phenomenological analysis while also drawing from the methodological practice of using individual differences to examine experience. These are similar to examples in the tradition of case studies where the analysis does not necessarily extended as generalizable to the whole population. The goal of this form of analysis is different. It allows for researchers to take seriously individual experiences as evident in the interview. To do this, participant interviews were analyzed, using the consensus categories discussed in Chapter 2. These analyses revealed that some participants had articulated varying degrees of awe and wonder (AW) experiences.

Here we look at specific examples. First, participant 14 (P14), a 20-year-old female in the FOC group, expressed varying levels of experience in multiple consensus categories: contentment (e.g. tranquility, relaxation), feeling overwhelmed, experiencing perspectival (spatial) change, bodily sensations of floating, and scale effects (e.g. vastness of universe, feeling of relative smallness). During the interview process, the interviewers employed open-ended questions to elicit these reports. The following interview excerpts provide examples of these expressions.

> *P14:* I think it was centered in on UCF and it comes out ... and ... I kind of like that feeling that it makes, I guess .... I don't know, I just like the way you feel when you feel like you are floating. ... I'm comparing the earth to the stars ... and how we are just this little planet around all these stars, like it's weird to me ... I guess just like how small the earth is compared to everything in the universe. I guess I was also thinking of like how different it looks looking into Earth compared to being on Earth and looking up ... just kind of uh, overwhelming, I guess ... cause it's, I don't know how exactly to describe it, it was just kind of surreal I guess, how small earth is compared to everything else .... The main thing that I was focusing on is, to me being on Earth it seems so big, but when you are really looking at Earth it's just, it's really small so it um ... it was just kind of like an awe moment type of thing – how small the earth really is and how I think everything is so big and important when really we're like the small little planet.

Participant 44 (P44), a nineteen-years-old female in the FOC condition, also indicated experiences that fit AW consensus categories.

> I was enjoying the different colors ... like each star had like a different color, some were blue and some were like a white color. Then I noticed some of the other blue ones were moving ... I just thought that they were really pretty [**aesthetic appreciation, pleasure**] and ... I guess I wondered if those were real stars or if they were just kind of a picture. I actually thought about the Hubble telescope once and wondered if this was like a real picture from like the Hubble telescope? ... I guess I was wondering where ... what was taking this picture and, like making the formation. [**interest/inquisitiveness**]
>
> It's kind of interesting to see because obviously you don't get that experience often because you're on Earth and so you're looking at Earth from being on Earth and walking around on it, but you don't really get that experience of looking down on it because very few people actually get to go into space so ... [**perspectival change**]

> It's almost overwhelming to just see everything you're experiencing, the stars and the water and the different continents all at once, and so just looking... [**overwhelming**]. You're [not just] looking at pictures and saying, oh, this is China and, oh, this is what the sun looks like; instead you see like all of it, all at once and you think, oh, this is what everything looks like put together...
>
> I guess when you see like a really pretty part of nature, like a waterfall or something. I guess, um, I kind of connect it to religion. I'm a Christian so I kind of connect it to God and how He's created these different places and He created the beauty, I guess, in your surroundings and stuff and there's just kind of a different perspective on the beauty that He's created, in my opinion... [**perspectival shift**] I feel like for a split second I thought, this must be like what God sees when He looks down on Earth, but I don't think I... I didn't linger on the God aspect of it, no.
>
> When everything is changing so quickly, you have so many thoughts all at once and you're kind of thinking about everything and you kind of, in a way, you get over... you get overwhelmed with thoughts, [**overwhelmed**] but then at the end when nothing's... everything's pretty much the same and it's just now zooming out, you just kind of relax and you just take in the full picture instead of just little things at once. [**contentment, unity of external**]

Not every participant shared these experiences. For example, participant 64 (P64), a twenty-year-old male in the FOC condition, reported nothing that the reviewers could categorize into any of the consensus categories. The same held for participant 65 (P65), a twenty-year-old female who did not indicate AW experience. Such a high discrepancy in the reports could be dismissed as purely behavioral (i.e. the AW "non-experiencers" could not or would not report awe and wonder), but, in that case, the physiological results should be similar to AW experiencers. This is not the case, however.

We analyzed these participants' reports, examining the individual differences of the AW experiencers (i.e. people who reported experiences of awe and wonder in the interviews) and the AW non-experiencers (i.e. those who reported no experience of awe or wonder). The latter participants certainly had their own experiences, they simply did not indicate that they had experiences that fit the consensus categories. We then looked at the participants' EEG difference from baseline (DFB) in average power spectral density (PSD) shown over the one-minute simulation time blocks. The power spectrum refers to the frequency and amplitude of each signal.

The frontal EEG sensors collected readings from the alpha, beta, and theta wave lengths. In frontal alpha (Figure 5.6), the AW experiencers (P14 & P44) showed greater suppression of frontal alpha than the AW non-experiencers (P64 & P65). The AW experiencers were both below the mean for frontal lobe DFB, whereas the AW non-experiencers had higher frontal alpha. The alpha readings were less distinct by experience over the central region (Figure 5.7). Alpha oscillations in the posterior regions (Figure 5.8) followed a similar pattern to those recorded from the frontal sensors with the experiencers showing consistently lower alpha in the occipital/parietal areas.

The alpha findings were similar when analyzed by hemisphere. Alpha in the left and right hemispheres was above the mean (and closer to

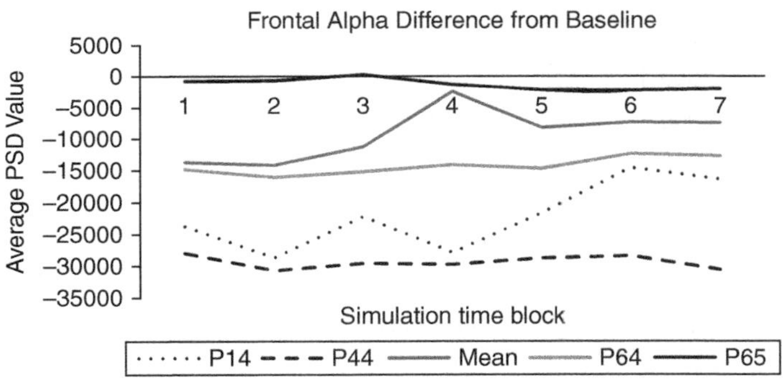

Figure 5.6 Individual differences examples for EEG frontal alpha

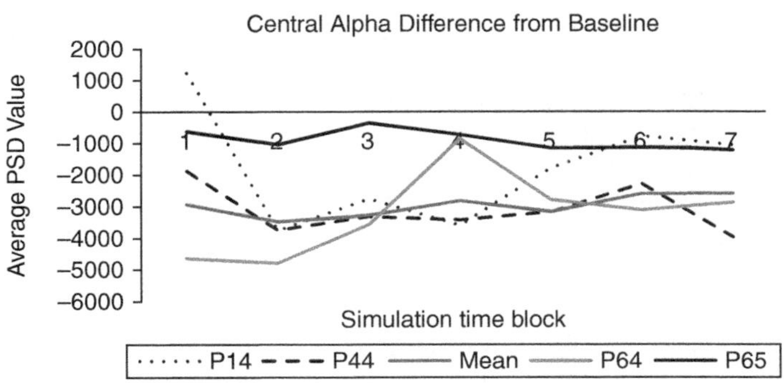

Figure 5.7 Individual differences for EEG central alpha

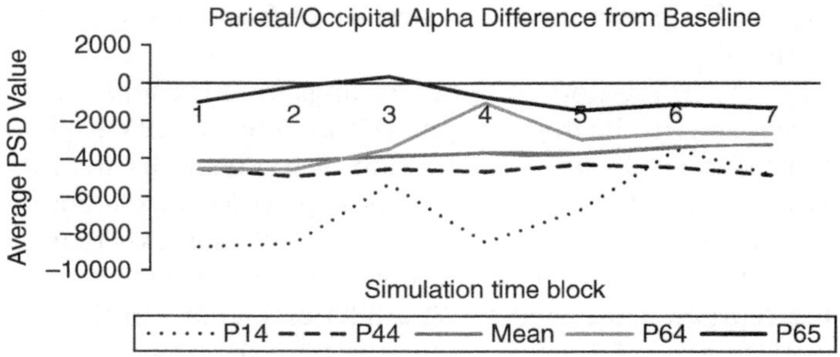

Figure 5.8 Individual differences for EEG parietal/occipital alpha

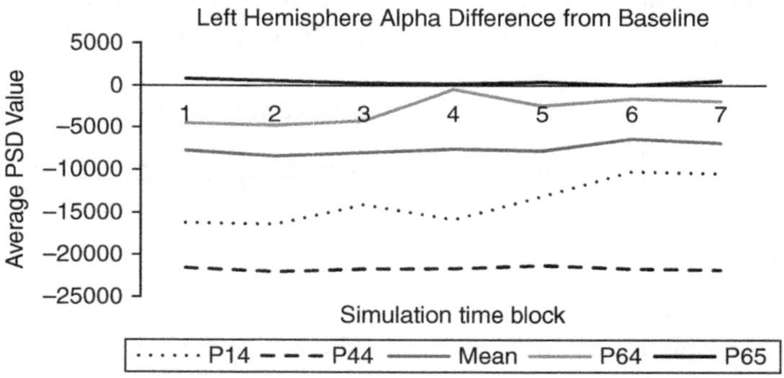

Figure 5.9 Individual differences for EEG left hemisphere alpha

baseline) for the AW non-experiencers and below the mean for the AW experiencers. Of note is P65 (an AW non-experiencer), whose alpha readings by hemisphere were statistically even with the baseline, and P44 (an AW experiencer) whose alpha stayed consistently below the baseline and mean by hemisphere. For the left hemisphere (Figure 5.9), P65 (M = 421.65) stayed statistically even with her baseline, whereas P44 (M = – 22,026.54) was below both her own baseline and the population mean (M = –7,748.91; SD = –11,515.40). Similar results were found in the right hemisphere (Figure 5.10), where P65 (M = –115.58) stayed statistically even with her baseline, whereas P44 (M = – 11,653.96) was below both her own baseline and the population mean (M = –5794.17; SD = –10,331.16).

For the beta and theta wavelengths, the differences from baseline were not as ordered, with the AW experiencers and AW non-experiencers showing no significant difference or consistency across sides of

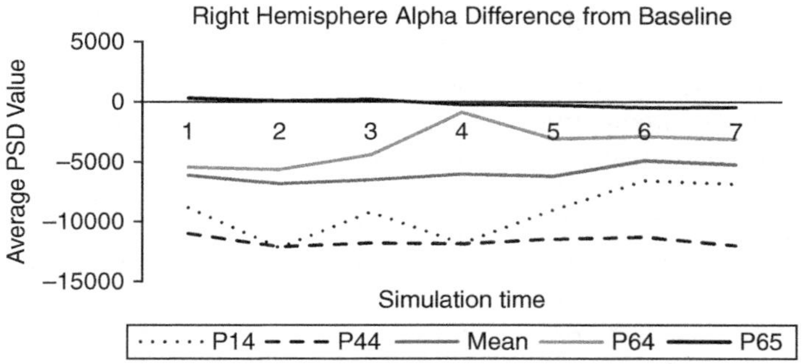

Figure 5.10 Individual differences for EEG right hemisphere alpha

the mean. These results suggest the functional utility of phenomenological data from the interviews in exploring individual differences for AW experiencers and AW non-experiencers.

3.3 Phenomenological and psychological results

The broad categories of AW experiencer and AW non-experiencer helped to identify participants for individual difference analyses described above. However, the consensus categories also lend themselves to a refined analysis of the psychological data as well. We conducted independent sample t-tests comparing AW experiencers and non-experiencers with responses to the questionnares. Significance was found for a direct grouping of AW experiencers and non-experiencers in relation to some factors in the ESSE, the TAS, and the BMMRS questionnaires. This comparison helps to identify relationships between the psychological data and the phenomenological data, thereby contributing to increasing the understanding of awe and wonder while validating the methodological changes. The following describes the results of the clustering of experience by broad categories of spiritual, religious, and aesthetic (see Chapter 2, Section 2 & 3) and the consensus categories.

Spiritual, religious, and aesthetic experiences

Participants who, during the phenomenological interview, articulated simulation-time experiences as spiritual (i.e., spoke of their experiences while viewing Earth in terms that indicate some aspect of self-transformation, e.g., perspectival (moral) shift) were less likely to describe themselves as "logical" in the ESSE; $t(44.759) = 3.435$, $p = .001$; spiritual-experiencers ($n = 45$, $M = 72.13$, $SE = 3.292$) < non-experiencers ($n = 16$, $M = 88.13$, $SE = 3.291$). There were no significant findings between the self-identifying logical category and any of the other psychological metrics from the ESSE, TAS, or BMMRS.

Due to the rigorous categorical limitations, religious-experiencers occurred less frequently than spiritual-experiencers, but when they did, these participants were less likely to self-report being a "reflective person" in the ESSE; $t(27.161) = -2.773$, $p = .010$; religious-experiencers ($n = 6$, $M = 84.67$, $SE = 2.472$),< non-experiencers ($n = 55$, $M = 73.64$, $SE =$ .3.116). These participants reported higher levels of curiosity in response to the simulation; $t\ (11.581) = -2.871$, $p = .015$, religious-experiencers ($M = 89.17$, $SE = 4.167$),> non-experiencers ($M = 74.31$, $SE = 3.069$). Participants who described their experiences in religious terms had higher rates of negative religious interactions in the BMMRS (Idler et al. 2003) than their counterparts who did not use religious language during the interviews; $t(21.423) = -2.486$, $p = .021$; religious-experiencers ($M =$ 7.83, SE = .167),< non-experiencers ($M = 7.22$, $SE = .183$).

Aesthetic-experiencers spoke of their experiences while viewing Earth in terms of the sublime, sensations, pleasure, and poetic expression. However, these participants scored significantly lower in the "sensory perceptual absorption" category of the TAS (Tellegen and Atkinson 1974); $t(59) = 2.292$, $p = .025$; aesthetic-experiencers ($n = 37$, $M = 14.38$, $SE = .407$), < non-experiencers ($n = 24$, $M = 15.83$, $SE = .477$). There were no other significant results of the t-tests for aesthetic experience.

Awe and wonder in the immersive simulation

After the phenomenological interview, participants who had expressed experiences of awe while in the simulation were categorized as "awe-experiencers" ($n = 39$) or non-experiencers ($n = 22$). The participants who reported experiences of awe during the phenomenological interview were significantly more likely to have reported awe in their psychological surveys as well, $t\ (34.018) = -2.374$, $p = .023$; Awe experiencers ($M = 19.69$, $SE = 3.626$) < non-experiencers ($M = 20.91$, $SE = 6.564$). While there was no significant relationship to wonder and curiosity, awe experiencers also reported greater levels of humility in their psychological surveys; $t\ (39.00) = -2.356$, $p = .024$; awe experiencers ($M = 63.74$, $SE = 4.843$) < non-experiencers ($M = 42.95$, $SE = 7.377$).

Participants who articulated awe experiences during their interviews also had higher scores in "private religious practice" in the BMMRS; $t\ (27.808) = -2.061$, $p = .049$; awe experiencers ($M = 21.77$, $SE = .514$), > non-experiencers ($M = 18.91$, $SE = 1.289$). Recall that the BMMRS is reverse scored, so higher scores mean lower levels of private religious practice. This is consistent with what we found in Experiment 1. No other area on the BMMRS showed significance with the interview expressions of awe.

Participants who reported experiences of awe in the phenomenological interview also scored significantly lower in the "sensory perceptual absorption" category of the TAS; $t(47.350) = 2.767$, $p = .008$; awe experiencers ($M = 14.33$, $SE = .477$) < non-experiencers ($M = 16.05$, $SE = .395$). The awe-experiencing participants were more likely to answer "false" to questions like, "Textures- such as wool, sand, and wood- sometimes remind me of colors and music," and "The crackle and flames of wood fire stimulate my imagination."

Participants who, during the interview, expressed experiences of wonder ($n = 26$) reported higher levels of awe in the ESSE; $t\ (58.910) = -2.022$, p = .048; wonder-experiencers ($M = 67.88$, $SE = 4.382$), > non-experiencers ($n = 35$, $M = 54.57$, $SE = .4.913$). There was no significant correlation between wonder-experiencers (as reported in the phenomenological interviews) and reports of wonder, curiosity, and humility in the ESSE, nor any of the other psychometrics.

4 Discussion

Our results provide several insights into the nature of awe, wonder, and some of their related constructs. First, however, we think it is clear that these phenomena and their interrelations are best identified through an analysis that considers the brain-body-environment as a dynamic system – that is, an analysis that considers the subject as in-the-world, and in these particular cases, as immersed in a virtual world. This becomes clear by examining the relationship between context and experience.

4.1 Context and experience

With significant results of EEG in the left hemispheric captured at various times and across all three wavelengths, a discussion regarding the role of the experimental manipulation should start with consideration of the left hemisphere itself. The left hemisphere refers to multiple regions of the brain left of the lateral fissure that are associated with a broad range of functions. The most useful approach, then, is to consider the functional associations most relevant to the phenomena under consideration. Perhaps, the most well-known contribution of the left hemisphere is its functional role in language (Vigneau et al. 2006). Both Wernicke's and Broca's areas are located in the left hemisphere, associated with both receptive and expressive language function in approximately 90% of all right-handed persons (Purves 2008). From a hermeneutical perspective, we know that language shapes experience, and although it is difficult to say how this works in regard to the particular experiences of awe and wonder, we find some important hints in the difficulty that some participants had in expressing precisely what these experiences were. From our theoretical vantage point, this is something that requires further exploration (see Chapter 6). It's clear, however, that experience is coupled to language and the capacity for expressing that experience. It's also possible that some LH change was involved in lexical activity (see below).

In addition to showing hemispheric shifts, the time segments were analyzed by specific wavelengths (alpha, beta, and theta). It is useful to consider what might be implied by these differences.

Alpha activity has been associated with a broad spectrum of conscious cortical activity, so there are numerous ways to explain significant variations found along this bandwidth. The following aims to disentangle some likely explanatory candidates.

Traditionally, alpha has been observed in "cortical idle" (Pfurtscheller and Lopes da Silva 1999), meaning alpha activity oscillates during alert awake states when one is not engaged in a task. In terms of the contextual changes that occurred in the simulation conditions, the greater change of left hemisphere alpha in the FOC group suggests integration of context and perception. The vehicle for the contextual integration may be a combination of lexical and embodied factors. Changes in alpha signal in the left hemisphere have been associated with lexical retrieval (Weems et al. 2004). During the phenomenological interviews, participants recalled engaging in impromptu "gamification" in response to specific stimuli while viewing the simulation, which may have been connected to lexical retrieval. The participants described trying to remember the names of landforms and bodies of water as the earth rotated.

The alpha levels in the FOC group may also be explained in relationship to unpleasant visually induced simulated movement (i.e., the rotation and lift simulated as the vantage moved from the earth to space) (de Toffol et al. 1990). The early presence of this effect might thus be attributed to the sense of dizziness that some participants reported when the simulation moved quickly over land. However, the discrepancy also appeared at the end of the experiment, when the visual stimuli for both FOC and GLO groups were quite similar. Viewing negative stimuli can also cause a depression in alpha (Makarchouk et al. 2011), possibly linked to limbic response. There was a drop in alpha in both FOC and GLO groups, with a greater drop in FOC, potentially indicating an unpleasant affective response to the grounding of the experiential context to the local campus starting point.

According to another interpretation, alpha differences at the beginning and end of the simulation may be related to changes in lateral gaze as the simulation moved from a centered focus (at the beginning of the FOC condition) to a full screen dispersed image (of earth landscape) to another centered focus (on planet Earth) in the center of the visual field (de Toffol et al. 1990). A final possible interpretation of these findings would be that the changes in alpha indicate shifts of task attention (Bonnefond and Jensen 2012). In considering the role of context, the view of the campus may have helped the FOC group generate and maintain attention. This interpretation has important implications for research on vigilance, as introducing contextual grounding into vigilance tasks may increase neurological attentive behaviors.

How then should we interpret the changes in Alpha? Here we start to see the importance of the phenomenological interviews. We need some indication from the subjects themselves about what the experience was like. In this respect, we get a better picture of the significance of changes in alpha by comparing AW experiencers (the data, for example, from P14 & P44) vs non-experiencers (the data from P64 & P65). Alpha in LH (difference from baseline 0) was above the mean for the AW non-experiencers and below the mean for the AW experiencers. Alpha oscillations in the posterior regions followed a similar pattern with the AW experiencers showing consistently lower alpha in the occipital/parietal areas. AW experiencers also showed greater suppression of frontal alpha than the AW non-experiencers. The AW experiencers were below the mean for frontal lobe difference from baseline, whereas the non-experiencers had higher frontal alpha. These changes in alpha thus correlate significantly with respect to AW experiences.

For AW experiencers vs non-experiencers, the beta wavelength differences from baseline were not as ordered as with Alpha. Global beta changes have been implicated in suppression of motor activity (Pogosyan et al. 2009). GLO participants had a significantly higher beta at both the beginning and the end of the simulation experience. GLO participants may have experienced a reduction of motor system response, perhaps due to there being fewer physical affordances presented within the GLO condition. The significant differences during the final minutes of the simulation are important, however. By the end of the simulation, the participants are viewing similar images with similar affordances (or the lack thereof). If the beta changes are indicators of interactive suppression, this also suggests that context has some influence on subsequent motor activation.

Theta differences pose interesting interpretive challenges for studies involving quiet contemplation, like the present study, as they are associated with both meditation and sleep. This ambiguity was one of the primary motivators for the methodological clarifications concerning experimental design. As mentioned in Chapter 3, the findings from Experiment 1 showed significant change in theta activity, but we could not conclusively say whether these were the results of relaxed and thoughtful states or transitions to sleep. We are not alone in struggling with the interpretation of theta. In some cases, left hemisphere theta reduction has been recorded during hypnosis (Taddei-Ferretti and Musio 1999) and suggests an increase in cognitive effort. Theta changes have been associated with meditative states, though studies conflict on the directionality of the changes for certain types of meditation (Cahn and Polich 2006). Experiment 2 reduced the length of simulation time, compared to Experiment 1, hoping to reduce the likelihood of drowsiness. Nonetheless, drowsiness may explain the similar theta findings between FOC and GLO groups in Experiment 2 as the simulation progressed, keeping in mind that for both conditions, the last few minutes were of a quiet, tranquil view of a slowly turning planet. The phenomenological interviews indicated that many participants felt relaxed, and even sleepy, by the end of the experiment. The simulation conditions were relaxing, with no audio stimulation, low lighting, and, according to most participants, pleasurable visuals. However, the significant difference between the two groups at the beginning was not likely caused by fatigue, but rather by the difference in first minute stimuli. The contributions of the other data sources (i.e. psychological and phenomenological) give a fuller experiential picture of relaxed, but alert states. Future work should disentangle the phases between thoughtful relaxation and drowsiness as it pertains to the first-person articulation of experience.

fNIR measurements also showed significant differences between the FOC and GLO conditions. Again, the key is to be found in the simulation timeline. In the first minute, significant differences are to be expected. The images are different, with the FOC containing various familiar images that, during the interviews, participants said they recognized. Many participants who received the FOC condition also reported looking for places, (e.g. trying to locate a girlfriend's apartment building or the route they take home). This type of engagement, or *gamification* could account for the differences in the frontal lobe behaviors, which are typically associated with executive function. Interviews from the GLO participants indicated a different sort of cognitive task, as they did not experience a familiar starting point. They started in darkness (similar to baseline, so it is not a surprise that there is less change from baseline here), and the first landscape images were not familiar. The vantage was over a red-toned landscape of Africa, and some participants reported thinking they were on Mars. The lack of familiarity at this stage may have made it more difficult to engage cognitively (Tulving et al. 1996). A similar issue of novelty versus familiarity may explain the differences during the fourth minute as well. However, this trend appeared throughout the experiment, even though the significance was only found during minutes one and four, suggesting an enduring effect of the initial contextual grounding on the subsequent frontal lobe behaviors.

Neurological responses to context differences between the FOC and GLO conditions indicate previously unexplored features of experience as it applies to the observation of Earth in a simulation environment. Consequently, there are implications for the experience of astronauts. These findings suggest that the grounded context, the notion of coming from "home" and moving into space, increases the neurological behaviors associated with both attention and relaxation. As the astronaut reports indicated experiences of peace and beauty, it is possible that the types of neural behaviors observed during the experiment are neural behaviors involved in transitioning from the anxiety of launch into a state that allows for more positive affective experiences while in space. Astronauts maintain a contextual awareness that they are leaving a specific location on Earth and they will return to a specific location.

Results from Experiment 2 suggest that contextual grounding is associated with differences in brain areas involved in attention, memory, and relaxation. However, while these findings begin to paint a picture of the neurological conditions associated with the experience of looking at Earth from space, they alone are not sufficient for describing the astronaut *experiences* of awe and wonder. Rather, it's the phenomenological interviews that suggest that the manipulation of one aspect of context alone is not enough to create or impede these experiences. To explore the nature of such experiences, these findings must be considered in their relationship to self-reports of the experiencers while viewing the simulation.

4.2 Complicating the phenomenology of awe and wonder

To flesh out the fuller figure of awe and wonder, it is essential to integrate the information collected from the participants in the form of their survey results. As the ESSE explicitly asked participants to report aspects of awe, wonder, and other related experiences, these reports were correlated to the neural results. Among the more intriguing findings are the opposing correlative directions for visual processing-associated beta and theta behaviors in "logical" versus "spiritual" or "religious" self-identifications. It seems that the self-identified spiritually inclined person not only sees the world differently *figuratively*, but quite *literally*. A person who considers herself "spiritual" or "religious" behaves differently on a neurophysiological level than a person who more strongly identifies as "logical." In interpreting these results, one must remember that the participants were free to identify with every, or no, category; that is, participants were not required to choose between "spiritual" or "logical" – they could choose both. Consequently, the self-identifications bound to neural behaviors are all the more intriguing.

It will be valuable to replicate these findings since they raise numerous questions for further study: Do other sensory modalities demonstrate such discrepancies (e.g. Does auditory processing vary in a similar pattern)? More broadly, what are the implications of such literally different worldviews for sociological and political progress? It is one thing to acknowledge that different cultures "see the world differently," but it is quite significant to suggest that our views of ourselves (our self-understandings) are closely bound to our sensory experiences. Future work should examine causality in this regard. Do I see the world differently because I am a spiritual person? Alternatively, am I a spiritual person, because I see the world differently? Perhaps, in such matters, traditional notions of causality begin to blur, and the exploration of this relationship opens possibilities for non-linear explanations.

More to the endeavor at hand, what does this self-perception say about the experiences of astronauts? One's self-identification as spiritual, religious, and/or logical is bound to one's personal history, inseparable from episodic memories, cultural differences, and the conceptual schemas we use for each construct. The neurological behaviors associated with experience are only *partially* the result of the stimulus and presented context. A large portion of the experience has to do with those things beyond the experimenter's control, the things unique to each individual. However, being beyond the experimenter's control does not mean that they are beyond the experimenter's grasp. That is where the phenomenological interview rounds out the toolkit of exploring experience. The tools of psychology and neuroscience tell us much, but they fall short of describing the experiences in the depth and fullness required. The phenomenological interviews support a broader sketch, an image of experience with dynamical movement, taking into account the complexities of individual differences.

In this case, the physiological and psychological findings are supported by the phenomenological interview, in that participants who self-identified as "logical" were significantly less likely to express themselves in spiritual terms. We want to suggest that this issue extends into the nature of the interview itself. If a participant who self-identifies as spiritual or logical gives an account from his or her personal perspective, in a way, the listener is exposed to a worldview that is more or less "spiritual" in experiential terms. Our own self-identification may act as a type of perceptual filter for ourselves *and for others*, and as such, the interactive aspect of understanding others may either facilitate or interfere with the framing of the other's experiential accounts (see Chapter 6).

Social signals, which include embodied aspects of posture, movement, gesture, facial expression, vocal intonation, but also communicative speech acts, can indicate something of intention, emotion, as well as personality, cultural background, and other markers of self-identity (Gallagher 2008; Gallagher and Varga 2014). Such things may give listeners information about a speaker's experience. Thus, if social signals indicate something about my interlocutor, then I may more accurately frame and understand an experience that the speaker is sharing with me because I have direct access to those social signals even though I don't have direct access to the experience they are describing. The self-described "logical" person may not only see things differently, but also in conversation, may invite the listener into a world that is shaped by that experiential difference. If the listener has not been exposed to the stimuli, then the shape of the speaker's perceptual filter may very well influence the second-person reception of the experience. The neurophenomenological approach contributes to a ground-level mapping of these difficult and entangled aspects of experience in an interactive world.

5 Conclusion

As the larger project of phenomenology seeks to describe and explore the nature and structure of experience, these distinct approaches to first-person accounts help researchers clarify the language used to represent and discuss the phenomena at hand. The results indicate that the consensus categories of awe and wonder experiences do cluster the participants into groups that coincide with predictable neural behaviors. In this case, the greater alpha-suppression rates in AW experiencers indicate the broader cortical activation required to synthesize consciousness, perception, and working memory. These categories, which originate in the astronauts' experience, can only be experimentally verified as a result of the voluntary, first person articulations of experience.

The phenomenological interview data thus begin to tell us something about the nature and structure of experience, particularly when analyzed in light of the psychometrics and neurophysiometrics. For example, people who articulated an experience of awe, were significantly more likely to have indicated experiencing humility on their psychological survey. A finding like this can provide important insight. Within the phenomenological interview, the articulation of awe experiences may not correlate to an articulation of humility, but that sense of humility may still have been present and is something reportable through the ESSE. In some cases, then, it is only through the analysis of the two data sources together that we begin to see a structural connection in the articulation of some constructs and the underlying experiences. These combinatorial analyses help to refine our interpretations, but more importantly, they direct researchers for future exploration to consider more closely the conditions under which these related phenomena co-occur.

Another example of combinatorial contributions comes from the integration of the TAS findings regarding sensory absorption and the phenomenological groupings. Contrary to what one might assume, aesthetic, awe, and humility experiencers scored significantly lower in sensory perceptual absorption. There are two possible explanations for this discrepancy, the first being methodological and the second, ontological. In regard to the methodological explanation, this may be a simple difference between what the metrics aim to measure and what they actually measure. This explanation could be explored by refining the metric, perhaps through isolating exclusively visual absorption (as opposed to multi-modal absorption) for a visually-exclusive stimulus. However, the fact that none of the other absorption categories were flagged for significance should elicit caution before dismissing the use of the metric. After all, the TAS category "nature and language" seems the right candidate for correlation when the items being discussed are articulations of the experience of viewing a natural phenomenon. Likewise, one might assume that a tendency toward "imaginative involvement" would rightly play a role in the experience of viewing a simulation. So, one must take seriously the second explanation – that the TAS is providing a proper measure of a distinction that may be hidden to phenomenological reflection. This suggests that the categorical relationships between these constructs should be more closely examined to parse out the structural commonalities for awe and wonder, as well as other spiritual, religious, and aesthetic constructs. We are, however, persuaded that the answers rely on a non-reductionist design and analysis of data that allow us to paint a more holistic picture of the phenomena.

The inquiry begins, and continues, anchored in the fact that actual experiences had by astronauts are accessible, describable, and subject to empirical investigation. It was with those experiences that the study began, using the hermeneutic analysis to frame the problem at hand. The visual component of the astronaut experience is only one component, the one mentioned explicitly by astronauts and something that can be simulated in a laboratory. However, by experimentally isolating that one controllable stimulus, it becomes possible to take a closer look at its relationship with the many factors individuals bring to their experiences. The results indicate that the nature of such experiences is not confined to astronauts in outer space. That is, the *nature*, the structures and tendencies of the experiences themselves, are like other experiences in that they are a function of dynamic interactions of cognitive, affective and physiological engagement in the world. In the case of astronauts, their prior experience is complex. The demands on astronaut bodies are remarkable, with extreme sensations in sound, touch, and sight. Yet, even with all of that extraordinary context, the ways in which the experiences of awe and wonder emerge can be considered as extensions of ordinary human experience.

With respect to the experiences that take shape in the dynamic relationships among histories, bodies, and environments, the present study, emphasizing the role of visual perception in relation to presented context, revealed phenomenological nuances of experience that would have gone unnoticed in traditional cognitive science. Subsequently, these relationships can be mapped to a model of experience, directly contributing to a non-reductionist picture and supporting the larger inquiry into the study of mind that will move us closer to understanding the nature of experiences, on Earth and beyond.

Part II

Insights and Extensions

6
The Phenomenology of Unprecedented Experience: Ontological and Cognitive Wonder

Is wonder inconceivable? The question is not, is wonder impossible, but is it inconceivable, that is, is it impossible to put into concepts, despite the fact that clearly people experience it, and we talk and write about it? Is wonder absolutely unique to each person who experiences it? Do concepts undermine wonder *as* wonder, turning it into something else that might be researchable, but is no longer the original experience?

These are questions which, in some more general form, have been at the root of social science methodology since the beginning. Human experience is always continuous, not discrete; embedded, not isolated; temporal, not eternal. It is always both nature and nurture, biological and social, caused and not caused, unique and yet prone to type or categorization. To the extent that we reduce experience to some category, we gain clarity at the expense of both comprehensiveness and relevance.

The experience of wonder is especially prone to reduction. It is easy to see it simply as one possible reaction to the unknown. Or, in a more sophisticated approach, it is easy to see wonder not as the sense of the unknown, but as the sudden recognition of order or meaning in what was previously disordered or lacking in meaning. Kant's wonder at the starry heavens above and the moral law within was surely his recognition of such an order. It is easy to simply rhapsodize about it, to just see the capacity for wonder as the source or the mark of our humanity. And, it is also easy to focus on what wonder accomplishes, rather than on the experience of wonder itself. This last inclination is particularly widespread among philosophers who, after Socrates, see philosophy as beginning in wonder. "True" questions (or as Gabriel Marcel might term them, mysteries) come from wonder, whereas problems arise from puzzles and have an end-point and clear solution.

All of these are forms of reduction, however, in the sense that the experience itself is not the focus. These tend to be about what wonder does, or causes, or leads to, rather than what the experience itself is. Husserl points us to a more interesting version of wonder. As Mark Kingwell puts it,

> What is wonderful, Husserl suggests, is not simply the oak leaf I look at, making me wonder why there is not nothing, for this feeling soon ceases. *What is also wonderful is this experience itself, and myself as the person in whom astonishment before the world is felt.* Wonder invites not only investigation of the world, but also reflection on the subject who experiences it, and on the experience itself. (Kingwell 2000, p. 89)

Husserl, Kingwell argues, moves us past the idea that wonder might be for something, that it is predominantly understandable inasmuch as it leads to other kinds of insight, or to philosophy, or to the divine or transcendental. Still, though, there is a tension in how we think of wonder that needs to be unpacked, and which is relevant to understanding some of the decisions made on this project, as well as decisions that might be made in future work.

1 Ontological and cognitive wonder

The version of wonder that is most common stems from the Greek word *thaumazein*. While the term is often associated with Plato's *Theatetus*, in which Socrates tells us that philosophy begins in wonder, the root of the term is in Greek mythology. Thaumas, who gives his name to *thaumazein*, was a sea god, son of Pontus and Gaia, married to Electra, and father of the Harpies and of Isis. As with all Greek mythology, the place of a god in a narrative and in an extended family is important. Thaumas embodies the wonders and dangers of the sea.

Thaumas is also a god of a sort of middle dynastic world. He is not of the chaotic age of Chronos and Gaia, and he is also not of the more orderly world of Zeus and Poseidon, much less the later age of the heroes depicted in Homer's *Iliad*. His world is not one of the randomness of chaos, but also not the well-ordered realm. *Thaumazein*, then, is not mere randomness, but is a realization of the hidden or transcendental order of things through an experience of finite beings. Thaumas's children indicate this duality – there are the Harpies, the destructive winds that snatch away things, and there is Iris, the rainbow messenger between gods and humans (and, incidentally, the part of the eye that delivers the outer world to the inner sense). She is the conciliator, also a wind but not destructive like the Harpies. Both she and her sisters deliver surprises, but of different kinds. The world is an ordered place, and wonders come from outside of the proper place of humans, outside of their place in the order. The ocean yields wonders precisely because it is not exactly a natural place for humans, and the same is true for the heavens – the stars and outer space.

This version of wonder has been handed down for hundreds of years, and the echoes of wonder as the recognition of a surprising order remains. This is not the only lineage of the concept, however. The English word "wonder" comes from the Latin *mirari* and *miraculum*, usually translated as anything wonderful or beyond human power, or an amazing event. There is in Latin a sense of incomprehensibility, even magic, about *miraculum*. These experiences astonished and amazed. *Miraculum* provokes "*admiratio*," a past participle of *mirari*. Our modern English derivation, "admiration," does not do justice to the level of astonishment implied in the Latin source.

Spinoza seems to present a secularized or naturalized version of the *miraculum* version of wonder. In *Ethics 3*, Proposition 52, Spinoza gives an enumeration of kinds of emotion, and considers whether wonder should be considered an emotion. He decides that it is not. "Wonder (*admiratio*) is the conception (*imaginatio*) of anything, wherein the mind comes to a stand, because the particular concept in question has no connection with other concepts." (Spinoza 1970, III.52 and note). Accordingly, he says, "I do not include wonder among the emotions [because] this distraction of the mind arises from no positive cause drawing away the mind from other objects, but merely from the absence of a cause." Wonder is seemingly "caused" by a lack of cause, but it in turn may cause other emotions such as consternation, veneration, horror, devotion, hatred, and so forth.

Spinoza is certainly not looking for miracles when he characterizes wonder in this manner; he is pointing to the cognitive and epistemological aspects of the experience. Whereas *thaumazein* may contain at least the promise of meaning, *admiratio* points to incomprehension. Spinoza's version of wonder recognizes a complete disconnect between one concept and all others. It does not require the recognition of some object or of one's place, it simply requires recognition of a lack.

It will remain to be seen whether these two approaches to wonder amount to the same thing in the end, but; for, the moment at least, they look different. We'll call the first, the Greek *thaumazein*, "ontological wonder" while the second, Spinoza's version derived from Latin, we'll call "cognitive wonder." The first emphasizes emotion (surprise) and perhaps revelation, while the second emphasizes concepts and their relation or lack thereof, and explicitly denies that wonder is emotion. The first is tied to action – wonder is not simply idle curiosity or puzzlement, but motivates reflective thought and is the basis for philosophy, while the second is produced by the mind's encounter with something that cannot be reduced to the realm of objects or causes, and may lead to emotion but need not motivate action. The first, ontological wonder, is tied to knowing our place in the cosmos and glimpsing something outside of that place (the sea, after all, was a source of mystery to the ancient Greek mind, since any kind of creature could emerge from it at any point). The second has nothing to do with our place, but rather has to do with the matrix of concepts. And, put most simply, philosophy may be based in *thaumazein*, or wonder, but is unlikely to be based in *miraculum*, or miracles.

One might suppose that, for phenomenology, this historical, philosophical background is beside the point. After all, phenomenology starts from experience, and what we have here is conceptual history. The problem, though, is that, even as we analyze experience through concepts, at some point there is the claim that these concepts add up to, or point toward, wonder. It is one thing to observe the experience of wonder phenomenologically; it is another to operationalize it within an experimental situation. Where precisely did the working definition of wonder come from? Just as Anselm's ontological proof does not, strictly speaking, prove the existence of God, but of "that greater than which none can be conceived," and he has to make the further step of saying, "this we call God," here too the set of concepts drawn from the astronauts' journals and reflections and noted in the experimental participants' interviews cannot in themselves point to wonder unless we know what wonder means, and accordingly have some sense of the conceptual history of wonder. Researchers, after all, did not abstractly test for wonder, or awe, but for the more concrete consensus categories, which are descriptive concepts. This historical sketch is significant, then, because when we think about how to study wonder, it is important to understand the kind of ontology that we are seeking.

2 Expressed and unexpressed experiences of wonder

Our project had, as its goal, to understand the experiences of astronauts during space travel, the experiences that were written about in their journals and other documents. This was to be accomplished by trying to model those experiences in a mixed-reality (Experiment 1) and an immersive visual environment (Experiment 2). The impetus for the study was the recognition that there was a family resemblance between reported experiences of those who had gone to space, from different countries and different times, that had not been adequately theorized, much less empirically studied. The project was premised on the idea that a set of categories could be extracted from the journals and reflections that would capture elements of the experience itself. No single concept was assumed to represent all the experiences; in fact, there were ultimately 34 consensus categories which were drawn from the writings themselves. Care was taken to work as closely as possible with the documents themselves, and precisely not to draw on philosophical tradition, the history of religious experience, psychological typology, or any other factors. The distinction between ontological and cognitive wonder was not assumed; in fact, wonder and awe themselves were deemed to be too vague as concepts at the analytical level, and so were not themselves directly used. Instead, the goal was to describe as carefully as possible the concepts used by the astronauts' to characterize their experience.

The methodology for working with these texts was a combination of experimental phenomenology and hermeneutics. Hermeneutics was useful because each word or set of words stood in an interpretive relationship with the concepts, as well as with the experience. That means that, while the concepts used were expressed in words, we did not assume that there was a one-to-one relationship between concept and word. In some cases, several different words might converge or point to a single concept, or, in other cases, one word could be used in different ways by different writers, thus indicating several different concepts. However, Hermeneutics is not merely a method for interpreting words,. There is always a measure of self-reflection in a properly applied hermeneutic method. It is always possible that the researchers may misunderstand the text, and, as such, the researchers' interpretations also stand as texts in need of interpretation. Because of this, the initial hermeneutical research done by the primary interpreters, Gallagher and Janz, was supplemented with checks and re-readings by 20 other people and groups acting as secondary interpreters.

The experimental phenomenology becomes relevant after the hermeneutical research has been done. After subjects went through the mixed reality or virtual reality simulations, they were interviewed by researchers trained in phenomenological interviewing techniques. These techniques involved open-ended questions that allowed subjects to describe their experiences during the simulation. The interviewers were trained to not prompt the subjects with any of the consensus categories, and the subjects were not aware of that list at any other time in the experimental process.

Here are two excerpts from the interviews. The relevant consensus category is identified in square brackets and is meant to identify the kinds of experience being discussed by the subjects.

Excerpt 1:

Experimenter (E):	What was it like to see the Earth mass in the simulation?
Subject (S):	That was, that was awesome. Um, it was just beautiful to see the water and the, like, landmasses and like, that large of a scale. And close to your planet, um, kind of like in front of you like you're above it or something. [Aesthetic appreciation]
E:	And what about it was awesome to you?
S:	Um, that just the detail I could see, like the water where it was rougher, the land, you could see where it was desert, where it was trees. Um, you know, it was just a different way of seeing Earth than I ever have. [Perspectival (spatial) change]
E:	And would that be similar to the "beautiful" that you described?
S:	Mhm.
E:	And are there any thoughts or feelings associated with the "beautiful" that you described?
S:	It made me want to stop littering (laughs). Like "save our planet!" (laughs) That'd be about it. (laughs). It made me like our planet more, I guess, or appreciate it, seeing how big and massive it was. [Scale effects; Perspectival (moral) shift]

In this excerpt, the subject expresses aesthetic appreciation of the scene, and draws moral implications from the experience.

Excerpt 2:

E: Okay. You've done a very good job of describing the physical sensation and your posture changes. What was it like inside, more on an emotional level, or a physical sensation level?

S: Well, it was ... I kind of ... I don't know, I was kind of like a little excited because it was just really cool to watch, like to see the whole earth like spinning around like in the orbit. I was just ... it was fun to watch, like I kind of pretended I was an astronaut, so that made it ... that made it kind of fun. So it was just ... I was sitting there just like watching the whole thing happen and I was ... I was kind of like going back to like my childhood where, you know, when you look on a map and you go, look, there's Florida, that's me, and I saw that. And I looked, oh, that's me. [Captured by View/Drawn to Phenomenon]

E: Okay. I see a smile in here too.

S: Yeah. ... when I was little I ... I mean, who didn't want to be an astronaut when they were little. That was probably like the coolest thing you could ever do. So I always, you know, I always like messed around with that. I always thought that it was really cool so when I saw ... when I saw like the whole simulation thing with like the Earth, I was ... I was watching it go around, I was like, oh, wow, this must be what it feels like to ... to really like see it from space.

One thing that is noteworthy in the second excerpt here, and is more apparent in the audio record, is that the subject stumbles (the ellipses indicate a pause or hesitation) in his/her attempt to describe the experience.

In the initial research methodology, the consensus categories that were derived from the astronauts' writings were used to interpret whether there was replication of the awe and wonder experiences connected with space travel in the simulation. The goals of the experiment hinged on the consensus categories. We wanted to see whether subjects, in the post-simulation phenomenological interviews, would describe their experiences using the same categories (which were never part of the previous communication with the subjects); likewise, these categories played a central role in guiding data analysis.

In all of our data collection, we assumed that the consensus categories and their related vocabulary would be what would yield positive data. And, this assumption is certainly warranted – replication of the categories would signal replication of the experiences, and that would give us significant data. But it may not be the only kind of significant data available in the interviews. The stumbling and the hesitation of the subject in this and other interviews may also be significant, precisely in the case of wonder, because wonder is so difficult to capture conceptually, in words. One necessary feature of it may be its ineffability. How might ineffability be captured? By looking for the gaps between concepts in the subject's reports, the places where the concepts do not easily exist. A gap in their vocalization, signified by an ellipsis in the transcript, could count as what the Greeks called, using the same term, *élleipsis*, which means, "omission" or "falling short." What's unsaid, in these contexts, may be telling us something. Is it reflecting something like Spinoza's notion of *admiratio*, or the Greek notion of *thaumazein?*

Ineffability, actually, may occur in either ontological or cognitive wonder, but for different reasons. If there is the perception of a larger order or reality for which there is no ready language, there might be gaps. On the other hand, if there is a concept that has no relation to other concepts, there might also be a lack of language to express this unique thing. In either case, stumbling and searching for words may be an indication that there is an experience worth attending to. In some cases, wonder may exist not in the concepts but in the gaps between concepts that are inadequate to the phenomenon.

3 Unprecedented experience

If the participants are struggling to articulate something that has no words, or perhaps no concepts, or is a concept with no connections to other concepts, a phenomenological study will have to be open to that possibility. There is work on the phenomenology of mystical experience (e.g., Steinbock 2007), but it does not explicitly suggest resources for, or approaches to, the NP experimental use of phenomenology. Furthermore, mystical experience often exists within a tradition, and experiencers seek it out, or at least prepare for it in various ways. While it involves unique experience, in other words, it may not be unprecedented experience in the mind of the experiencer in the same sense as the experience of wonder in our astronauts. Furthermore, there has been very little scientific study of mystical experience (as opposed to, for instance, contemplative experience and mindfulness, which has been the subject of extensive scientific research (e.g., Didonna 2009; Wallace 2007).

The phenomenology of mystical experience, however, can help us focus on experience itself, as opposed to much of the work on religious experience that does not bracket off claims about transcendental causes, or assumes substantive universality in the experiences. Another similarity between wonder and mystical experience, which goes back at least to William James, is that mystical experience has been seen as ineffable. So, while there may be a tradition and preparation for many mystics, there may still be some sense of inexpressibility about the experience, and hence some sense of unprecedentedness. Phenomenology, then, at least takes us to the first step by bracketing any metaphysical and epistemological claims about experience.

We are still faced with the problem, though, because much of phenomenology assumes experience to be relatively widely distributed, widely enough so that phenomenologists themselves can share in the experience. The point here is not empirical generalizability, but the *apriori* notion that there is a universal structure to experience.

Wonder may, in fact, be a category we use for a set or range of different *sui generis* experiences. Phenomenology does not spend a lot of time thinking about the question of, for example, whether the blue that I see is the same as the blue that someone else sees – one of the metaphysical assumptions that is bracketed is the assumption of the priority of individuality, or the solipsism of experience. Blueness exists in a world of experience that is not just mine or yours, but ours, and the experience of blue is tied to the experience of communicating about blue, accomplishing things in the world where we assume that we agree on what blue is, and so forth. Blueness, in other words, is neither just about wavelengths of light, nor is it just a *quale* (that is, an individual mental experience); rather, it exists in a meaningful and shared world.

And, of course, so does wonder. Moreover, when it comes to wonder, the cognitive content of the experience includes affective states and other reported elements that can be quite similar across different instances. This can lead us to believe that wonder is like blueness, that it fits into a world of meaning. But with wonder, it is at least possible that every instance of it is *sui generis*. In other words, affective and other aspects of experience may be meaningful, but also ineffable, and fundamentally different for everyone who experiences them. In fact, to suppose that wonder fits into a world of meaning can incline us in the opposite direction from the one just described, that is, towards the assumption that reports about affective and other aspect of experience are all about a unified thing – an experience of wonder that is the same for all of us.

In researching mysticism, this has led people in the past to argue for what is known as a "perennialist" approach, which suggests that all mystical experiences are of the same thing, and that further they undergird religion and other meaningful experience. Many continue to hold this kind of perennialism, but it is important to realize that there is nothing in the experience itself, even in its extended, social form, that requires experiences of wonder to fit into the same meaningful world. Your wonder may not, in fact, be my wonder at all, and the wonder experienced by some astronauts may be quite different from that of others.

The earlier distinction between ontological and cognitive wonder becomes relevant here. Ontological wonder tends toward a unified account of wonder – there is order out there, apprehended but not understood by us, which produces the wonder effect in us. Cognitive wonder, on the other hand, is a concept with no connection to other concepts. There is nothing necessarily unified there.

4 The categories of wonder

One issue with using phenomenological method to study awe and wonder concerns operationalization. In this case, the word is understood as the move from the explication of experience to the use of that experience to structure the neurophenomenological experiment. At one level, we sidestepped this problem by running three kinds of analysis in parallel (phenomenological, psychological, and biometric) and then looking for correlations between them. Still, the move from the experience of the astronauts to something that could be included in tests and then finally flagged in phenomenological interviews required some elements of that experience to be identified and compared across these moments. We used the consensus categories as these elements – they are available in both writing and speaking, they are one means of framing issues of meaning and its lack, and they are (or can be) elements of experience itself, as opposed to just reflection on experience. They capture, perhaps, and may be able to capture *only* the idea of ontological wonder.

Accordingly, there are potential limitations of working with such categories. For one, concepts can tend to privilege the intellectual over the embodied. In our project, that issue is mitigated somewhat by the relational methodology the researchers used. Affective and embodied aspects of the experience were not ignored in the use of the consensus categories; they were articulated in the post-simulation interviews. And, the other aspects of the experiment (the psychological and biometric) kept issues of embodiment central to the study.

Another possible issue with using the consensus categories to study wonder is identified by Anthony Steinbock in his focus on "verticality" in his phenomenology of religious experience. He frames awe and wonder as follows:

> What is given vertically incites awe, and only later as a consequence, wonder. Modes of givenness [i.e., experience] are "vertical" in the

sense that they take us *beyond* ourselves. These modes of vertical givenness are testimony to the radical presence of "absolutes" *within* the field of human experience. (Steinbock 2007, pp. 14–15).

For Steinbock, verticality adds a dimension to phenomenology that the focus on horizontal givenness (i.e., experiences that can be captured in categories) has largely forgotten. While he argues that it can be found in the later work of the phenomenologist Merleau-Ponty, among others, it is worth noting that, for Merleau-Ponty the vertical is pre-spiritual, although that may still mean that it is something that transcends categorization (Merleau-Ponty 1968, p. 204).

Wonder, on Steinbock's view, is a later consequence of verticality (although, again, whether it is for Merleau-Ponty is unclear). Steinbock treats it as a weakened version of awe, or mystical experience (in a sense, not so differently from the justifications of mysticism found in an earlier generation, in people like Evelyn Underhill (1912) and Friedrich von Hugel (1908), as the foundation of religion, morality and so forth). Various other experiences, such as experiences of epiphany, revelation, manifestation, and so forth, are also seen as forms of vertical givenness. Although he does not explicitly do so, Steinbock might argue that focusing on the consensus categories in relation to wonder effectively substitutes "horizontal" phenomenological concerns for vertical ones.

This issue is important for our analysis, and Steinbock's phenomenology of mystical experience stands as the most sophisticated version of this approach in a long time. If we are to see awe and wonder as ersatz forms of religiosity, there might be a case for seeing the framing of awe and wonder through consensus categories as missing the point, as factoring out the very verticality that makes the experiences meaningful. If, however, we do not start from the idea that awe and wonder are simply poor echoes of religiosity and see them instead as experiences in their own right, it is possible to regard the list of consensus categories as a useful phenomenological tool. Furthermore, proceeding as we have done, does not preclude also seeing experiences of wonder as nascent religious experiences. For the purposes of our study, however, with one exception, the notion of religiosity *per se* was bracketed unless directly referred to by a text or interview.

The one exception involved the results from the BMMRS questionnaire. The questionnaire explicitly measured aspects of religiosity and indicated that awe and wonder experiencers showed, in some respects, less or weaker indicators of religiosity than non-experiencers. If Steinbock is correct and phenomena such as awe and wonder are indications of religiosity, we would expect a positive correlation. One possible explanation for this, (if we frame religiosity in terms of traditional religious or theological beliefs), is that such beliefs might in some cases dampen any experiences of awe and wonder rather than enhance them. If the experiences come with a closely correlated pre-conceived theological belief or explanation, that belief or explanation may impose a kind of regulation that shifts the energy away from the experience itself. Arthur C. Clarke (1973) is famous for his "three laws of prediction," one of which was "any sufficiently advanced technology is indistinguishable from magic" (1973, pp. 14, 21, 36). In other words, anything that does not have an apparent explanation can be the source of wonder. Thus, if there is an internalized explanation in the form of strong beliefs about divine agency, wonder may cease being quite as wonderful. It may simply be perceived as another confirming instance of an already held belief. As we indicated, this is also consistent with Keltner and Haidt's (2003) model in which awe and wonder experiences do not depend on positive religiosity and may be connected with the inability to incorporate an experience into one's conceptual schema. That is, subjects who indicate that faith is comforting and who engage in religious practice may be better able to accommodate the effects of the visuals or may have different expectations that modulate what they experience as awe and wonder.

5 What can we say about the unsaid?

The final issue involved with using consensus categories has already been raised. It is that the concepts or categories can make discrete what is not really discrete. Furthermore, they can tend to privilege what we have words for. The experience of wonder may be the kind of experience which is meaningful but not easily articulable, or in fact completely ineffable. The ineffability may be of the sort that fades over time due to increased familiarity with the conditions of the experience, or it may be robust, remaining inexpressible despite familiarity with causal mechanisms or correlative factors. Cognitive wonder, especially of Spinoza's sort, would tend to the first, as the "concept with no connections to other concepts" may nonetheless start to collect connections with other concepts. Ontological wonder, on the other hand, may have the potential to remain ineffable – if it is ineffable to start with. The wonder that philosophy begins in may well be seen as renewable, resistant to being resolved or explained like a problem rather than a mystery.

This is perhaps the most serious question to be raised about the use of categories to study the experience of wonder. What if the experience of wonder is inherently resistant to conceptualization, and the act of looking for categories of such experience actually undermines the possibility of considering the experience of wonder itself?

We should keep in mind, however, that the consensus categories were derived from the experiences described by the astronauts. If these space visitors are unable to capture everything about such experiences, the consensus categories obviously capture something about some experiences that were shared by the astronauts during space flight. These are not predetermined concepts and categories that we are imposing on experience from the outside. They are, we could say, bottom-up, coming from first-order experience, rather than top-down, imposed on first-order experience. Not all experience, however, is conceptual. Indeed, we tried to reflect that in our working definition of wonder: *A reflective feeling one has when unable to put things back into a familiar conceptual framework*. Can there be a category of experience that falls outside of any category? Perhaps that's just what wonder is.

Yet, the ontological version of wonder suggests an experience of a kind of order, as opposed to the experience of an object in isolation, and order seems close to the idea that something falls under a concept or category. Plato, and in a different manner Heidegger (1992), see wonder not in mundane objects themselves but in a certain order that pertains to or reveals such objects – the experience of forms, or the encounter with *alethia* (an order of disclosedness), shows reality as it is, and produces wonder. Likewise, the wonder that happens for astronauts in space is quite literally not an experience of mundane (worldly) objects, and indeed seems to be unprecedented experience from which emerges a recognition of previously unrecognized order. The wonder that comes with the recognition of order may be like the experience of "faciality" (Kozin 2007). We are inclined to see faces even in things that are not faces – it is a kind of organizing ability that we are cognitively disposed toward. Husserl, Levinas, Agamben, Sartre, Merleau-Ponty and others use this sense of recognition to ground intersubjectivity, ethics, politics, and a host of other things, but for our purposes here, faciality suggests a capacity for experiencing or being affected by meaningful order. It is not a surprise that the wonder that people have while gazing into space is compared to looking into the face of God. As Astronaut Edgar Mitchell put it,

> It is a universal feeling among astronauts and cosmonauts. It's what I call instant global consciousness.... Mystics would call it a mystical experience, psychologists a peak experience, theologians would say it

> was touching the face of God.... My view of our Planet was a Glimpse of Divinity. (Mitchell, Apollo 14, in Osborne 1997, pp. 10ff.)

Another astronaut, Peggy Whitson, expresses it in less religious, more aesthetic terms as she describes crossing the line that separates day from night.

> [On the horizon] the sun shows a blinding face that burns the atmosphere with molten reds and oranges before seemingly melting itself into the darkness, leaving a royal blue line that dissipates more slowly as the stars come out from hiding. (Peggy Whitson, http://www.nasa.gov/mission_pages/ station/expeditions/ expedition16/journal_peggy_whitson_7.html)

We should think of the consensus categories then, as approximations, not of the objects of wonder but of the experience of wonder; not of the face, but of an experiential faciality. One experiences insight as one grasps an order that had not been grasped before; one experiences a shift to a perspective that one had not occupied before; one experiences a certain unity that had not been previously disclosed. It's an experience of the unfamiliar, of feeling not quite at home with what confronts us, motivating perhaps a nostalgia, which may also be seeing one's home from a different perspective. It is also clear that those approximations have as much significance for what they do not capture as for what they do capture.

If we approach the issue by assuming that wonder is cognitive, we can see how this might work. A concept that does not connect to other concepts will, at best, be understood affectively as unfamiliar, and perhaps can only be approached by means of implication and extrapolation. Understood this way, the use of categories of experience in the project makes sense. The categories are ways of making wonder concrete, ways that capture whatever they can capture, and in doing so, allow it to be studied. Categories make sense in regard to the ontological version of wonder as well. If wonder shows forth an order which we grasp perhaps only dimly, an order that can surprise and overwhelm us, that order may be expressible in a category, or may be expressible only in the failed attempt to express it, in the ellipses that signal an impulse to search for a category yet unsaid.

Husserl argues that *thaumazein* is what distinguishes the "universal but mytho-practical" attitudes of other world philosophical traditions from the "theoretical" tradition in the West. While his Eurocentrism is problematic and without anthropological foundation, his point about wonder is the same as Plato's – *thaumazein* leads to theory. Classically, *theoria* was contemplation, for Plato, contemplation of the forms, and for early Christian theology, contemplation of God. Theory was accomplished through the formation of concepts or categories – in the logic of the time, these were the only ways of accessing unchanging reality. Theory for Husserl refers to a particularly human, non-reductionist attitude towards the changing world, one which is shown to be universal when metaphysical assumptions are bracketed and experience is understood in a disinterested manner. As such, the theoretical attitude is the basis for science.

The effect of space travel in regard to the experience of wonder may well be that it simply makes it more available. If we normally find ourselves in what Husserl (1970) calls the "natural attitude," a mundane attitude concerned with practical matters, it is hard for wonder to break through. In space, where the unprecedented is closer to a type of encounter we have with another person's face, the membrane between the natural pragmatic sense of how things work and the theoretical attitude is thinner. Wonder has a chance to show through again and may not simply be mistaken for spectacle or novelty.

If it is the case that wonder has multiple historical trajectories and capturing the related categories requires also capturing the spaces between the categories, those places where subjects struggle to articulate their meaningful experience, future research will need to find ways to recognize that kind of complexity. An experience of wonder is, in some significant sense, unprecedented for the one experiencing it, and while it might be a universal among humans (enough that it can ground philosophy and perhaps even science), it cannot be approached as a universal at the level of the primary or first-order experience itself.

7

Imaging and Imagining Space: How Popular Culture Shapes Our Expectations of Outer Space

> If you could see the earth illuminated when you were in a place as dark as night, it would look to you more splendid than the moon. – Galileo Galilei, *Dialogue Concerning the Two Chief World Systems*, 1632.

It is remarkable to think that visual phenomena that we take for granted today, and that seem self-evident and in no need of further explanation, appear in history only at a specific point in time. Today it is possible, for example, to create and reproduce photographic images of lightning flashes and the Milky Way. In regard to such images, the usual thought is that an image simply represents what already exists. In this chapter, we look into this commonly held belief, because in the history of images this simple relationship between a phenomenon and its representation is called into question by a theoretical perspective that claims the contrary, namely, that images can and do, to some extent, also have an impact on that which is being represented. The fact that people did not always depict phenomena, such as lightning or the Milky Way, and facts about the way they began to do so, allow inferences about their perception of these things (Baigrie 1996; Daston and Galison 2007; Galsion and Jones 1998).

1 Some examples from before the space age

We want to address these general considerations within the history of images using representations of outer space, *from* outer space, that is, from a perspective that very few people have actually had. It is thus useful to initially remain on Earth and look at the perception and visual depiction of things that our readers presumably have already seen, beginning with ones close by. Humans can only visually reproduce things that they can actually see. The invisible can naturally remain concealed in a number of ways. The human eye may not be able to see an object because it is too distant, too small, or moving too fast, or simply because it is concealed or hidden. The human exploratory urge led, by at least the seventeenth century, to the use of ground lenses to develop microscopes and telescopes, which disclosed visual worlds previously concealed for lack of these necessary aids. A further massive visualization impulse then came in the nineteenth century with the development of photography, which, used serially, provided an early form of film.

A very clear example of a visual phenomenon that had not always been visible stems from this time. It concerned a dispute about the exact motion of a horse's legs during a gallop. As work animals, horses had been closely observed over many centuries. People were well acquainted with the characteristics and movements of horses, but they were unable to accurately observe the galloping gait because of its high speed. The numerous pictures of horses seen throughout art history (since antiquity) usually show the animal in the pose of an equestrian statue as an accolade to its rider or as a weapon on the battlefield. With the advent of sport and thus of horse racing, the horse and rider, also at high speed, became more interesting as a visual theme. Predestined for this new pictorial task in the early nineteenth century was the talented painter and jockey Théodore Géricault, who had gained prominence with his famous 1819 painting of *The Raft of Medusa* and was in search of new imagery for his brush. His double set of skills drew him quite naturally to the theme of horse racing. So, in 1821, he painted *The Epsom Derby*, now in the Louvre in Paris. This painting is of interest in the history of images, because it reveals how the painter and horse enthusiast Géricault – an expert, therefore, in both visual representation and in that being represented – saw the horses' galloping gait. Moving at full speed, the horses simultaneously stretch their front legs out ahead and their back ones to the rear. Although this detail of the painting was not uncontested, many equine experts also "saw" the horse's legs in this position during a full gallop. At the time Géricault created his painting, photography had not yet been invented, and the exposure times in the imaging technology's early period were nonetheless too long to capture something lasting only a fraction of a second. The exact motion of a galloping horse's legs would thus remain hidden from human view for some time. Eadweard Muybridge's (1830–1904) serial photographs were, therefore, a sensation (Figure 7.1). In 1872, he became the first person

Figure 7.1 Eadweard Muybridge's serial photographs, The Horse in Motion, 1878

to photograph a horse at full gallop in a sequence of photographs (taken quickly one after the other), and then to display these images in quick succession, creating the effect of a film. This visual sequential depiction of the motion showed unambiguously that Géricault's earlier assumption was wrong, that a horse's front and rear legs never simultaneously extended away from its body during a gallop.

A simple faith in technology might have led us to believe that Muybridge's apparatus had given us sight. Undoubtedly, it did. However, there are two other aspects that offer further insight. First, the "knowledge" is quickly also transferred to the experience of seeing – performed without visual aids. Anyone who views Muybridge's film, *Horse in Motion,* will thereafter also see every live horse's galloping gait differently. The film's evidence influences the act of seeing (Barber 2012). We do not have the feeling of having understood something, but rather the feeling – and this is the crucial point – of seeing something directly, and in some very real sense, we do see it directly, when before we couldn't see it at all. The second point relates to this: this new way of seeing a motion – which some years earlier had been invisible because of its speed – now quickly began to affect the broader popular culture. With the popularity of horses as toys, or in cartoons, or adventure films – all present in huge numbers of children's rooms in the Western world of the twentieth and twenty-first centuries, literally every child can now not only cognitively describe, but also see a horse's gallop "correctly."

We chose this example as an introduction to what we want to say about the shaping of our perceptions of outer space by popular culture, because it provides the right picture, namely that the following is concerned, not with intellectual knowledge, but with a perceptual knowledge imparted by seeing and informing our imagination.

In regard to the representation of outer space, here too, the concern is not with the retrieval of knowledge about outer space, or galaxies, or planetary systems, but rather with how we imagine outer space to be. On Earth, we are closest to outer space when we look at the sky, and typically we think of it as looking at the night sky. Outer space is for us a visual. Celestial phenomena, in turn, offer good examples of the influence of depictions on our perception. Let us remain on Earth for the moment and imagine two situations. The first is a night-time thunderstorm with lightning, the second a starry night allowing a view of the Milky Way.

Staying with the first situation, imagine a bolt of lightning. Seeing a bolt of lightning is an experience that everyone has certainly had. Lightning was, of course, depicted in the medium of paintings before photography existed. Storms with lighting were a common pictorial theme in the seventeenth century. The natural night is usually not as dark as depicted in photographs, but it allows flashes of lightning descending from the clouds to the earth in their typical zigzag form to be seen quite clearly. Thus, there is no reason why lighting flashes should not always have been an object of the history of images. Although a flash of lightning moves very quickly to earth and represents a very sudden event, this example, in contrast to that of the galloping horse, can be seen with the naked eye and without optical devices. Lightning was thus already being represented in seventeenth century painting in a manner similar to modern day.

However, it is surprising that flashes of lightning are absent in the Italian Renaissance, a period positively obsessed with the sky, stars, and astronomy. Paintings showing thunderstorms do indeed exist, as here in Giorgione's *La Tempesta* (Figure 7.2), but the flash of lightning is rendered quite differently (Colin 2002; Rieth 1953). In the middle of the painting, we see warm light, similar to sunlight, penetrating a fissure in the clouds. The light does not have the typical zigzag form and the bolt has no connection to the earth or other flashes of lightning.

It remains an oddity that painters in periods boasting a skillful and naturalistic representation of most objects, such as landscapes, people, clothing, and many other things, could not depict other phenomena, such as a flash of lightning. As Reith (1952, p. 25) suggests of the lightening in Giorgione's, *La Tempesta*, it's a pantheistic representation. Or did the artists actually paint the flash as they saw it? Or was it just that they saw the natural flash of lightening differently? This is a difficult

Figure 7.2 Giorgione's *La Tempesta*

and far-reaching question, which probably cannot be answered here. We mention it because images of outer space possibly occupy a special position in regard to this question. We will come back to this issue later.

Before that, consider one more example: the Milky Way. Certainly, the fact that some celestial phenomena were not "seen" correctly was not due to a lack of attention. Numerous broadsheets have been preserved showing people keenly observing the sky, with the celestial phenomena described in text beneath the image. These sheets generally show the stars distributed evenly across the sky. What we do not see, for example, is the Milky Way. And here, too, the following question emerges: When did people begin to see the Milky Way as a bright band stretched across the sky? People in the fifteenth century who quite intensively observed the sky and disseminated their findings in images should have seen the Milky Way. But here, too, we make the surprising discovery that our galaxy appeared comparatively late in images.

In the history of images, Adam Elsheimer's 1609 painting, *Ruhe auf der Flucht nach Ägypten,* is considered the first example of a depiction of the Milky Way. The viewer sees a nighttime scene imbedded in a religious storyline, the flight to Egypt. The galaxy's band of stars stretches from the upper left to the lower right. It is known from sources where and when this painting was created, allowing researchers to compare the painted sky with a computer-aided situational image of the sky over Rome on June 19, 1609 (Hartl and Sicka 2005). The comparison showed that Elsheimer, too, did not reproduce an exact, one-to-one image of the sky. Nonetheless, he was the first person to depict the Milky Way, and he was probably one of the first artists to use the newly invented telescope to observe the sky and especially the moon.

2 Exploring outer space

The effect described in the previous section concerning celestial phenomena in the history of images can be amplified even more when applied to the experience of outer space. It is not only true that, so far, very few people have been able to have that experience, but also that this experience had been simulated for a long time prior to our space journeys. In popular culture, humankind traversed galaxies well before the first manned space journey in 1961 (see Geppert 2012).

The first blossoming of these space travel fantasies came in the first half of the twentieth century and can thus claim to be a special case in everyday culture: Space travel was already popular even before it was possible.

One genre of popular culture, namely science fiction, as manifested in particular in literature, film, and comic books, developed during the early twentieth century and continued to flourish beyond 1969, when the first human stepped onto the moon. As a rule, a person, object, or activity acquires "pop" status after achieving widespread attention and acceptance, in other words, after a long process and using channels of mass publicity. It is not only a matter of public taste, but also the involvement of industrial forms of distribution and the power of the media. In the case of space travel, stories and visions circulated long before anything had occurred in reality.

In the case of the galloping horse, people had already seen a real galloping horse; however, only by seeing the images of it did we learn to see how it really galloped. In contrast, in the case of space travel, before ever having actually seen outer space from the perspective of being in outer space, we were learning how to see outer space from images.

Here, the relationship between visually experiencing the actual phenomenon, the knowledge gained in the process, and the popular processing in visual media appears to have been reversed. The desire to reach the stars was so strong that it was already the subject of the very early film, *Le Voyage dans la Lune*, made by Georges Méliès in 1902 (Douy and Douy 1993). Admittedly, this film was still entirely a product of a fantasy world, but what we want to show is that our basic conception of what outer space looks like actually arose from early science fiction representations and remained in effect, even as the proportion of fantasy diminished, in the years following, in favor of a type of future-oriented documentation. In this regard, we will discuss our preoccupations with the moon, and then turn towards the Earth.

2.1 Focus on the moon

The moon has always been a presence in the night sky, and it is therefore not surprising that the burgeoning pioneer spirit of nineteenth and twentieth-century industrialization, punctuated by developments in steam engine technology, submarine capacity, and mechanical flight, also carried technological fantasy into outer space. Popular culture reflected the dream. The film, *Destination Moon* (see Figure 7.3), released in 1950, became the first great postwar cinematic classic. Not only is space travel depicted in this film prior to the first space flight, but a moon landing as well. It contains scenes of weightlessness outside of the spaceship, with the film's emotional highpoint certainly being the moon landing and the ensuing shot panning across the lunar landscape.

Although the film's astronauts do view the earth with amazement in the key scene, they then turn their attention to their true objective: the moon. In this film sequence, the earth as viewed from outer space is initially shown, albeit as a crescent, the shape we on earth most commonly see when viewing the moon, rather than as the familiar full globe "Blue

Figure 7.3 *Destination Moon* (1950)

Marble" image that we are used to seeing. The camera, however, immediately turns its focus from the earth to the moon. The earth plays only a minor part and is given little priority as a point of visual interest. Unquestionably, the moon has the "starring role" in this film. The earth is treated simply as a home port, and intervening space, merely as the zone that must be traversed in order to reach the destination.

Needless to say, the television as a technological object was also part and parcel of the enthusiasm for technology that characterized the 1950s; so it is hardly surprising that an early television series titled *Captain Video and the Video Rangers* (1949–1955), and the related film serial, *Captain Video: Master of the Stratosphere* (1951) dealt with the subject of space travel (Weinstein 2002). The series drew on a rather simplistic opposition between good and evil. It nonetheless continues to be of special interest to art historians because of its early handling of new or unknown technologies, including a robot named Tobor. Its protagonist, Captain Video, had the ability to see through all materials, thanks to his arsenal of technical gadgetry. And naturally, not only is he visually omniscient; he is also able to fly. The malevolent alien, dressed in a costume reminiscent of a fierce Viking, reigns over his own technical arsenal; he can, for example, accelerate the trajectories of asteroids and deploy them as weapons against Captain Video. Such effects are presented using interpolated animated sequences.

On the whole, the series is remarkable for its vigorous "anything goes" attitude and the infatuation with technology, so typical of the 1950s. The protagonists are never astonished to find themselves in outer space, a situation they regard as no more exceptional than that of riding a bicycle.

Just a few years later, this attitude made possible a film that no longer settled for narrating fiction, but which instead would raise certain documentary claims: *Man and the Moon* was a Disneyland television series episode that originally aired on December 28, 1955. Directed by Disney animator, Ward Kimball, it begins with a humorous look at humankind's fascination with the moon. This animated segment features lunar references ranging from the writings of William Shakespeare and children's nursery rhymes to lunar superstitions and scientific research. Kimball then makes an appearance to talk about the moon using informational graphics, after which he introduces Dr. Wernher von Braun, who discusses plans for a trip around the moon. Disney employed Von Braun and a number of other people as technical consultants on this film (Newell 2013).

The film's entire composition reflects the idea that space travel as a serious subject evolves out of a popular set of images, indeed, sometimes humorously animated film. This transition is evident also in the case of Wernher von Braun, who came from Germany's World War II arms industry to become a leading figure in the American manned space flight program (Biddle 2009). Wernher von Braun's appearance in the film marks the end of lightheartedness and a transition to seriousness. This move to earnest matters is implemented visually in the film's setting. Standing in front of an illustration of the first phase of a trip to the moon, an orbit of the rocket ship around the earth, Von Braun explains the fundamental possibilities of an expedition to the moon. With cinematic skill, the researcher presents his vision – not as a distant future happening, but as an unavoidable next step for space travel.

The film then shows a live action simulation from inside and outside the manned Lunar Recon Ship RM-1 to illustrate how such an expedition might actually occur. (This included a scene of a near disaster as a very small meteor punctures one of the spaceship's fuel tanks). This staged emergency is employed less as a means to show that the planners had thought of everything, but rather as a cinematic trick to illustrate that the film episode was presenting, not only an ideal series of events but also the handling of unexpected events. The prominent role of the moon mentioned earlier is also accentuated visually in this episode, with the object of desire visible in almost every scene.

The film's success was proved by its lasting popularity. It was re-aired in 1959 under the new title *Tomorrow the Moon*. These early films were marked by a dispassionate enthusiasm for technology and by functional precision. This attitude was mandatory for producing this type of "future documentary," which, rather than being fiction, was a kind of anticipatory reportage.

During this period, the moon was the object of limitless visual desire. All eyes were turned toward her, and it was only later, with the early, unmanned flights, and the later, manned moon voyages, that the earth entered into the viewer's field of vision (Byers 1977). But here as well, the demands of the earth's inhabitants contributed to shaping its subsequent image.

2.2 Focus on the Earth

On first glance, it seems logical that the focus during the development of the popular images of outer space would initially be on the destination, with the view being turned back toward the home planet only after that goal had been reached. This change of viewing direction, however, was also connected with a change in mindset. The years leading up to the first space launches were shaped by an enthusiasm for technology and accompanied by a firm belief in progress. Such a mentality is based on the conviction that only one direction exists: forward! Once an objective has been achieved, the next one is established and pursued. A look back is, at most, a trace of sentimentality, in which a person striving for the future briefly indulges before pushing forward even more resolutely. This kind of scientific drive was evidenced in *Destination Moon*. A new element emerges as the earth becomes the focus of visual representation: knowledge about our planet that was unavailable here on earth and could only won from a viewing point in outer space. It is not solely the case of an image fulfilling expectations, but sometimes that of expectations demanding an image, which then changes and shapes our idea of outer space.

The so-called Space Race between the United States and the Soviet Union commenced in the 1950s (Neger and Soucek 2011). The first man-made spacecraft to reach the moon was a Soviet probe Luna 2, which impacted on September 13, 1959. The Luna 9 was the first space vehicle to land softly on the moon, touching down on February 3, 1966. The first humans to walk on the moon were the Apollo 11 astronauts, who landed there on July 21, 1969, at 3:56 CET. They were followed by five subsequent manned moon landings as the Apollo Program continued over the next three years (Mayer 2011).

In the course of this development, it was not 1969, that is, not the year of the actual moon landing, but rather 1966 that was of special significance in the history of images. This was not necessarily due to any particular image, but by virtue of the demand for an image. By 1966, the "spaceship earth" metaphor had entered the public consciousness. In September of that year, the United States Vice President, Hubert Humphrey suggested that "the earth itself is a kind of manned spaceship."

> As we begin to comprehend that the earth itself is a kind of manned spaceship hurtling through the infinity of space – it will seem increasingly absurd that we have not better organized the life of the human family. (Speech at San Fernando Valley State College, 26 September 1966; cited at http://www.spacequotations.com/earth.html)

That same year, the economist, Barbara Ward, published *Spaceship Earth* (1968), launching the idea of sustainable development. Stewart Brand, who later became editor of the *Whole Earth Catalog,* put out a call for a particular image.

> It was February 1966 and I was twenty-eight and was sitting on a gravelly roof in San Francisco's North Beach. I had taken a mild dose of LSD

> on an otherwise boring afternoon and sat, wrapped in a blanket, gazing at the San Francisco skyline. As I stared at the city's high-rises, I realized they were not really parallel, but diverged slightly at the top because of the curve of the earth. I started thinking that the curve of the earth must be more dramatic the higher one went. I could see that it was curved, think it, and finally feel it. I imagined going farther and farther into orbit and soon realized that the sight of the entire planet, seen at once, would be quite dramatic and would make a point that Buckminster Fuller was always ranting about: that people act as if the earth is flat, when in reality it is spherical and extremely finite, and until we learn to treat it as a finite thing, we will never get civilization right. I herded my trembling thoughts together as the winds blew and time passed. And I figured a photograph – a color photograph – would help make that happen. There it would be for all to see, the earth complete, tiny, adrift, and no one would ever perceive things the same way. (Brand 2009; cited at http://en.wikipedia.org/wiki/Stewart_Brand)

Brand initiated a campaign to get NASA to publish a photo of the whole earth, with some success. NASA generated a single black-and-white photograph of the earth on August 23, 1966 (Figure 7.4).

The photograph shows a close-up view of the moon surface in its lower portion, with the distant Earth above, approximately half in shadow. The view back towards Earth is very similar in composition to the one already known from the 1950 movie, *Destination Moon*. In comparison with the color version from that science fiction film, NASA's black and white photograph fell short of expectations. This photograph was, in fact, the first of the earth in its entirety taken from moon orbit. The image, however, could not achieve any major degree of impact, despite

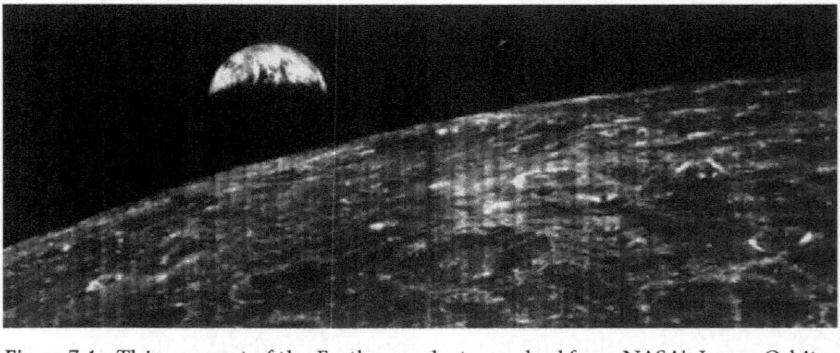

Figure 7.4 This crescent of the Earth was photographed from NASA's Lunar Orbiter I, August 23, 1966 when the spacecraft was just about to pass behind the moon.

its unquestionable superlative nature, because the earth's inhabitants already knew the perspective from popular culture. Reality could only catch up with, but not outdo, what science fiction had already broached. This "familiar" image could only fulfill Stewart Brand's demands on an objective level. When the visionary, Brand, called for a photograph of the earth as a whole, he naturally did not mean a half-shadowed view. A complete view of the earth, however, was published a year later.

This was the first color photograph of the earth (as a whole) taken by the third Applications Technology Satellite (ATS-III) on November 10, 1967 from a point in geostationary orbit above Brazil. South America is seen quite clearly, as is the west coast of Africa in the upper right section of the image and part of North America in its upper left. The ATS-III was launched on November 6, 1967, and is still in orbit. The communications system of the ATS-III was used to support the Apollo moon landings in 1969–1972.

Yet, this image was also unable to earn a place of honor in the history of images. It's a good example that demonstrates that images can only succeed in popular culture if they satisfy the expectations people have of them. When Steward Brand formulated his demand for a view of the earth, he probably had only a vague notion of this yearned-for image, a notion preconditioned by fictional images found in popular culture. A "real" image of outer space could first fulfill these expectations when both aspects – the actual photograph taken in outer space and the one characterized by popular culture's expectations – had been similarly realized.

The image that eventually fulfilled these criteria would become one of the most well known photographs of all time and a symbol of an entire generation. It was taken, presumably by astronaut Jack Schmitt or Ron Evans, with a 70 mm Hasselblad camera and an 80 mm lens, on December 7, 1972, from the Apollo 17 space capsule at an altitude of 29,000 km. Beginning with the first moon flight in July of 1967, the earth as a whole was photographed repeatedly. The picture of the blue planet that would become the quintessential image of Earth known under the title "Blue Marble" (Figure 7.5) was not recorded until the Apollo 17 flight in 1972 (López-Alegría 2014; Wilkinson et al. 1998).

The photograph originated at a time when space photos enjoyed a high level of credibility. But the previously mentioned photographs could also make this claim to veracity. In contrast, the Blue Marble shot with its view from space bore witness to an almost metaphysical beauty. It revealed, for the first time, that the earth was a globe of seeming perfect harmony.

Figure 7.5 The "Blue Marble" photograph, taken on December 7, 1972, by the crew of the Apollo 17 spacecraft at a distance of about 29,000 kilometres. It shows Africa, Antarctica, and the Arabian Peninsula. Public domain image. http://www.nasa.gov/images/content/115334main_image_feature_329_ys_full.jpg

The original photograph was reversed top to bottom, but it was rotated for easier orientation. The reversal maintained the convention that north is "up," and allowed viewers to imagine that they were actually flying through space while looking back at Earth in a standing position (similar to how Brand had imagined and requested it in 1966).

The impression of perfection only strengthened the accompanying message of vulnerability. The dark sky surrounding a solitary Earth awash in blue appears to be a background din of danger. The interplay of beauty and danger produced by the distanced viewpoint had an unprecedented effect on humankind's emotional temperament. With its view of Africa and showing large sections of Antarctica through the clouds, the Blue Marble presented two of the earth's particularly burdened problem areas, thus concentrating in this view in particular the dual aspect of beauty and threat.

The photograph was a very powerful symbol of the change of perspective seen in the 1970s, from exploitation of mother Earth to caring for the planet as for one's own mother. The image was often distributed within this context of feelings and insight bearing the caption, "Love your Mother."

The Blue Marble's suggestive power ensured the re-establishment of the metaphor of the Earth as a living being. The Gaia Project, initiated by former NASA consultant, James Lovelock (2000), has become particularly well known. Lovelock found the image of the shimmering globe against the black infinitude of space so compelling that it led him to develop a fundamentally new understanding of the earth. He was of the impression that Earth's delicately formed biosphere, with its respiration-like motion and functions, belonged to a living orb. With the name 'Gaia' Lovelock invoked Hesiod's *Theogomy*, which describes a personified Mother Earth rising from the chaos and receiving the name Gaia. The development to a pantheistic concept of faith meant that it was not far to an occult Gaia cult, which the New Age movement ascribed to the Blue Marble (Bredekamp 2011).

This photograph also became an icon of the ecology movement and an emblem of our planet's vulnerability. The earth's centrality in the image has distracted from the fact that it does not, as a rule, look like this from space. Nonetheless, the Blue Marble is simply our image of Earth – it's what we think the earth should look like from outer space. It conforms to an ideal.

The expectations attached to photographs of outer space so evident in the case of the iconic Blue Marble image eventually also began retroactively to affect less prominent images made available by the NASA

Figure 7.6 The Earth–Moon System (http://solarsystem.nasa.gov/multimedia/gallery/Earth_Moon_br.jpg)

databank. In this regard, consider a photograph showing the moon as well as the earth (Figure 7.6). The photograph "Earth-Moon System" also offers insight into the processing of images in accordance with specific expectations. In the NASA databank, it is published with the following annotation: "Contrast and color have been computer-enhanced for both objects to improve visibility." The question, of course, is: What type of visibility is referred to here? We suggest that the visibility was improved to meet popular expectations. Ultimately, these expectations as an aspect of a history of style will also change.

3 In the simulations

In the Virtual Space Lab of the Experiment 1, we found that expectations of the participants were shaped more strongly by cultural images, especially films they had seen, than in the second experiment. We think this is because we primed their imaginations with the space flight narrative, the "suiting up," the launch sequence, and the physical layout of the VSL, with its out-of-window views. Even the content of the simulation, especially the visual entry of the ISS into one of the simulations, motivated references to popular culture. We launched the participants into a culturally pre-defined space in Experiment 1 more so than we did in Experiment 2. Experiment 2 promoted a different kind of immersion – immersion in the purely visual aspects of space – and seemingly, although this type of immersion generated experiences of awe and wonder, it did not explicitly elicit, as much as the simulation in Experiment 1, aspects that directly related to or reflected popular culture.

3.1 Cultural interventions in Experiment 1

In Experiment 1, of the 42 interviews conducted, 15 of them (35%) contained explicit references to movies. These included specific mentions of movies such as *Apollo 13*, *Armageddon*, and *Star Wars*, or to television series such as *Star Trek*. In some cases, the movies or shows framed the experiences the participants were having in the VSL, creating expectations of what was to come next. In other cases, participants entered into what we might call movie mode, thinking of themselves as if in a movie. An example of this latter can be found in the interview with Participant 4 who explained that he supplemented his simulation experience by imagining himself "flying in a space shuttle, where everything is like.... It was just kind of a visual... kind of movie. Just really short. Very short movie... just me flying around in space and... the Earth is on my left and the moon is on my right. So, I'm flying between them." Participant 18 adopted the film metaphor of "panning out" during the simulation to describe what he was experiencing when he experienced scale effects.

Participant 41 also associated scale effects with a movie. He spent the first two minutes of the simulation thinking about astronauts. When, during the interview, he is asked what he was thinking about during that time, he responded : "Kind of like movies and stuff like the *Apollo 13* movie with Tom Hanks. That's when I started thinking about it a little bit.... I started thinking [about] the distance where you could put your thumb over the earth. That's what Tom Hanks did in the movie.... It was basically about that scene where he puts his thumb covering the earth and I thought it was really cool because you're just putting your thumb over a huge population of people, and that was basically what I thought."

The use of popular film to frame the participant's experience can be seen in Participant 21 who indicated that Star Wars and other movies helped him during the simulation to think about the view of the earth, having seen "this shot before" in movies. Indeed, this framing seems to interrupt his experience of awe – "[Awe and beauty] – not enough to take your breath away or anything but it still makes you stop. For a moment your mind is not wandering. You don't want to think about anything else, you're just looking at [what is] in front of you." Then when the mind does start wandering, "I [started] thinking about the astronauts, seeing how they would see things and then a clip of a movie comes to my head and other movies. That probably triggered it [i.e., the mind-wandering]." Asked to say more about this, Participant 21 states: "I started thinking about different movies where I had seen astronauts [involved in] different things going on [in the] spaceship or whatever. I was thinking about [different films]: *Armageddon*, was one of them, *Apollo 13* ..." Participant 21 makes it clear that visuals from those films are passing through his imagination: "a couple of different [scenes], some when they're floating around like they had just gotten up there and they can see outside the window for the first time.... [In] one movie ... *Day After Tomorrow* ... the astronauts are like oh it's clearing up for the first time, I was thinking about that part." He confirms that these are all visual imaginings.

The phrase "clearing up" is a reference to cloud cover. The experience of Participant 26 makes this clear. P26 talks about his expectations of seeing the earth without clouds, and how these expectations were shaped by movies. But what he saw in the simulation, which included cloud cover, wasn't "normal" – that is, wasn't what he expected. The interviewer asks him how precisely he thinks of the globe without clouds. "Do you have a certain image in mind, a memory, or do you say something to yourself?" P26 answers: "Like in movies, I always see the earth from space it's like it's so normal. It [what he actually saw in the simulation] wasn't what I expected.... At first it was the clouds because I couldn't really distinguish any continents; [just] more clouds."

Participant 29, like P26, mentions her expectation of seeing Earth without clouds – but then notes all the clouds. Interviewer: "The window opens, tell me what you see."

> P26: "It was that...if you had seen the movies or the internet I suppose, it's just that picture of earth that you always think of, [but] you see the clouds. I remember looking at the clouds on the earth and seeing [them] as it zoomed out..."

Participant 33 tries to frame his own experience with reference to a film and, as he says, to "things I've seen even in movies where they are space flying and they're looking at the earth and that sort of thing. I was just trying to figure out if that's what I'm looking at." Looking at the earth, thinking of maps, he is reminded of movies where they are doing similar things. Likewise, P19 refers to the movie *Armageddon* to helps him imagine some of the personal problems astronauts might have when they are so far from the earth. He thinks about astronauts looking for home but all they can see is cloud cover.

> P19: "I thought back to the movie, *Armageddon*.... I kind of thought like the earth is so far away. It's like a whole new world out there. I thought back to how [the astronauts, like the characters in the movie] would feel."

Similar framings of experience by movies can be found in other interviews. Participant 6 thinks of the movie *Avatar* when he starts to see the earth come into view, and he starts to imagine that it could be a different planet from earth or a place like Pandora in the movie.

Participant 13 started to think about space movies to help himself conceive of being in space or living in space and being able to see satellites. Participant 18 references the movie *Thor* in order to help him think of multiple universes. "One of the things that I did think about that was kind of odd; I thought earlier about the earth orbiting and stuff like that. I saw *Thor* a few weeks ago and how [there were] different universes I guess. So I thought about that a little bit. Not so much the movie but just how there's different universes and stuff like that." Earlier in the VSL Participant 18 had been reminded of *Apollo 13* simply by sounds he heard "when they [the experimenters] were hooking everything up and then you could hear [voices] in the background. I wasn't really sure what the sounds were but it sounded like, I guess, if you were in one of the control rooms and it sounded like *Apollo 13* when they're all talking to each other in the control room. That's what I thought of."

In one of the conditions in Experiment 1's simulation, the ISS enters the scene. This particular condition elicited further references to movies. Participant 32, for example, is reminded of a movie – a Disney movie – when he sees the space station appear – specifically a movie where people live on the space station. "I thought about a movie, it's like an old Disney movie, where people live in the International Space Station sometime in the future and when I was a little kid I thought that that was the whole purpose of the International Space Station." Likewise for Participant 27, the appearance of the space station makes him think of a movie, which he can't remember, except for a scene that resembles what he sees in the simulation– and it prompted him to look to see if there were any people on the IIS. If it was the same Disney movie, it was likely an expectation shaped by the movie scene. "This space station thing, I didn't know what it was... It starts from the left side of the screen, crossing over to the right side and it was big."

> Interviewer: "Ok tell me what it looks like."
>
> P27: "I forget. I saw that in a movie once though – like rectangular pieces kind of stuck together – I can't really explain it.... It just slowly starts crossing from the left side to the right side. I was observing the space station thing. I guess [I was looking at] the whole entire structure of it. Like how it was [structured] and if there are any people on it maybe.... There were no people."

During the simulation P27 spends about 20 seconds trying to recollect the space ship and "what it was doing in a movie that I watched... and I kept going back to watching it."

> Interviewer: "Ok so this thing looks like something you'd seen in a movie?"
>
> P27: "Yeah."
>
> I: "When did you make a connection with the movie image and this image?"
>
> P27: "Right when the image started getting more into the screen because at first all I saw was a straight line like a rectangle, like a piece

> of it so I didn't know what it was and then when I saw the middle part of the structure, I knew."
>
> I: "Ok when you recognized it and you think back to this movie, how did you recollect the movie?"
>
> P: "I don't remember. [It was] something that I watched when I was younger and I just remember that scene."
>
> I: "Ok and that scene that you remember is that an image?"
>
> P: "Yes."
>
> I: "Ok and when [during the simulation] you think back to the movie that you saw when you were younger, how did you feel?"
>
> P: "Like normal."

Participant 34 associates the appearance of the ISS with another space ship, this time in a Star Trek episode. He anticipates the appearance of a space ship. The ISS appears. He then introduces a bit of action.

> P34: "Have you seen *Star Trek* or any episodes of *Star Trek*? ... I'm a pretty avid *Star Trek* reader and fan .... I was thinking that it would be cool if I saw the Enterprise or some other spacecraft fly by or something. Even though I didn't know if that was going to happen or anything because satellite technology isn't advanced enough to be in the same [league with] science fiction .... Just when I was zooming out from the earth and I saw – before I even saw it I was thinking about seeing a spacecraft coming across the screen and then I saw the satellite and I was thinking of *Star Trek* still."
>
> I: "Ok what happens next after the satellite appears?"
>
> P34: "Well I can see it moving across the screen. I'm just looking at it. It looks like the Hubble Satellite to me. I can see the solar panels. I made out the solar panels on the satellite. As far as my imagination goes I was still thinking about *Star Trek*. I thought it would be cool if the satellite blew up."

Perhaps the most dramatic set of unfulfilled expectations sparked by the VSL simulation and recall of either movie scenes or video games is found in the reports of several participants who thought they might encounter aliens. Participant 5: "At first it was kind of exciting when the window opened, and I was like, "Yes! What's gonna go on?" I was hoping I'd see aliens or something. But, I didn't see anything. It was just the stars...the stars aren't bad." And later he says, "I was kind of disappointed though...Cause nothing else happened...It's just...We moved out and stopped...No aliens. No meteorites...No nothing...Just the Earth." Likewise, Participant 24 specifies what he had been anticipating as he looked out of the porthole waiting for something to happen: "Something random like an alien or something; or [waiting for] an explosion [to] happen." Similarly, Participant 34, the serious *Star Trek* fan: "I was sitting there and anxiously waiting to see what's going to happen and then this simulation began and at first I thought it was going to be pretty intense but then I realized after a while I was just observing celestial objects and then the moon and then the earth. But at first [I thought] aliens were going to pop out or something or that I was going to get attacked or something because I was under the assumption that I was in a military simulation. So ...but then I realized that wasn't the case. I just relaxed and was just observing."

Since our experiments were conducted at the University of Central Florida in Orlando, a good number of our participants were familiar with simulations and simulated rides at popular local theme parks, like Disney World and Universal. At the very beginning of the VSL simulation – the launch sequence – Participant 9 thought about a ride he experienced at Disney World. "It's like one of those rides at Disney, say you're at Disney [on] that take off ride and you normally feel it shaking and [it] makes that same exact noise of taking off. So if you were to close your eyes, and actually get into it, then it feels like you're really taking off."

3.2 Theme parks, video games, and movies in Experiment 2

The influence of Disney also found its way into Experiment 2's simulation. Participant 16, in Experiment 2, highlighted the differences between our simulation and Disney's virtual adventure ride *Mission to Mars*.

> P16: "I've been to Disney a few times, so I thought the space [simulation] reminded me of *Mission to Mars*, which is like the modernistic spaceship one. Other than that I guess that's the only thing that reminded me of it."
>
> I: "Ok and what kind of emotions and feelings do you have when you think about *Mission to Mars*?"
>
> P16: "I mean that one's exciting, its like a lot more fast paced, like there's a lot more going on; this [one] was...very like at a slow pace and the same thing goes on the whole time."

Disney was also on Participant 33's mind when she entered the simulation. "I was honestly a little nervous coming in, cause I don't know how I felt about um.....like I don't know if you've heard of *Mission Space* at Disney or whatever. A lot of people say they get like motion sickness, at first I was a little nervous, I wasn't sure what to expect, but then it was fine." Likewise, Participant 10 compared the experiment's simulation to the Spider Man ride at Universal in that both made him feel dizzy.

Two participants in Experiment 2 referenced video games as part of their experiential descriptions. Participant 40 described the simulation as it rotated and zoomed out. "It didn't stay fixed in a North/South orientation.... It started out as what looked like a satellite image, at first, but it just became like a computer rendering...and I have to admit, I sometimes play a space simulation video game, and I know as we hit a certain altitude for it to, umm, start its gravity turn and begin accelerating into orbit [short laugh], so that was one of my – that was an expectation that was broken. Umm, so we didn't do the gravity turn and we continued straight into orbit." And Participant 50 compares what he's seeing to Google earth and then says: "It looked kind of similar, probably a little more detailed. Umm, it was slightly similar, so I was like uh, this reminds me...I play a lot of video games, so it just reminds me of computer-generated things. So it was just hard for me to picture that [image] as the real Earth. So that, I think that was the main issue." Participant 50 continued: "I think it was some time during the beginning [of the space simulation], the whole beginning to the zoom out sequence. I was imagining myself playing video games on it....Umm, but, to describe it, I think I would, you know...you could really imagine yourself looking at the earth from a rocketship."

To the extent that thinking of video games or theme park rides led to disappointed expectations, it pushed a few of the participants into reflections on the technology and in that way interfered with the immersion experience.

During the second experiment, there was significantly less references to movies (only 5 out of 63 participants mentioned movies). Participant 7 described his own zooming in on aspects of the simulation as a kind of zoom effect one finds in movies. Participant 53 indicated that the beginning of the space simulation reminded him of the end of a movie: "it reminded me of you know how like at the end of movies produced by Universal, they have the Universal [logo] with the planet earth and it rotates. That was the first thing that came into my mind and I felt like I was watching the beginning of a movie." Participatant 63 thinks in terms of movies he's seen in regard to looking at different parts of the earth – i.e., the movies in which he saw these countries – e.g. *Indiana*

Jones, or the TV series *Lost*. Two other participants used movies to frame their thoughts about people in different countries. Participant 32, when he thinks about impoverished people in Africa, he thinks of movie depictions of these people and starts to feel sad, guilty, etc. Likewise, Participant 58, reflecting similar feelings, thinks of the film *Slumdog Millionaire* when he sees India.

It seems quite clear, then, in both experiments, but especially Experiment 1, for many of the participants, movies were in the background shaping some aspects of what they were experiencing, seeing, or expecting to see. The films, and other aspects of popular culture, may have presented a contrast with what they were currently experiencing in the simulation. Alternatively, such things may have reinforced or in some way supplemented what they were seeing. In some cases, especially those of contrast, when their expectations were disappointed, these things interfered with the effect of the simulation, and prevented or disrupted experiences of awe and wonder.

4 A final peek into the depths

We conclude by discussing a work of art devoted to the heavens: the *Sterne* (Stars) photographic series by the German photographer, Thomas Ruff (Bono 2011). More than any other contemporary artist, Thomas Ruff, who studied under Bernd and Hilla Becher, has reflected on the links between photography, its technological prerequisites, and the demands society makes on the medium. Ruff has examined these interconnections in a number of photographic series. Even as a child, Thomas Ruff was fascinated by astronomy, and the universe plays a prominent role in these series, as do the images we humans make of the universe.

Ruff produced his first astronomical series, *Sterne,* between 1989 and 1992. He lacked the technology needed to realize his vision of the stars and their appearance, so he chose to use scientific images available from the archives of the European Southern Observatory (ESO). Six hundred six 29 x 29 cm negatives taken with a telescope in the Chilean Andes form the basis of this work, with the resulting photographs encompassing the entire sky of the southern hemisphere. The artist enlarged selected, coherently composed details from each negative in a portrait format with a 5-degree angle of view. For each image, he sought to use the technology and format that would best communicate his conception of it. It was a priority to produce "not a copy, but a picture." The vertical portrait format allows the viewer's gaze to range further into the distance, as though looking up into the sky. Ruff divided the selected images into six groups and gave them titles such as "1. Photos of foreground stars with normal star density in the background" and "5. Photos of stars with interstellar objects and gas nebulae." He did this to highlight the scientific origins and encyclopedic ambitions of the source material.

This artistic approach resulted in a series of large-format photographs showing a sky filled with stars – views which neither the human eye, nor a quick look through a telescope would normally reveal in this format. Although the images are effectively unaltered scientific photographs, the recorded objects – distant stars, galaxies, and gas nebulae visible against the black background – are at times extremely dim and difficult to make out when viewed with the naked eye, so Ruff has in some ways created an illusion, a cliché of a starry sky. Through his work, Ruff expresses the idea that expectations held of supposedly scientific images can influence their form to some extent.

This concept could also be confirmed on a historical level. A final example, also of an artistic nature, shows that this mechanism applies not only to an ever more distant region of the galaxy, but also to areas of outer space actually considered unreachable from today's perspective.

In her series, *Night Sky* (2007–2008), Angela Bulloch used commercially available software to calculate views of outer space not available to the inhabitants of Earth (Mühling 2008). With the use of LED technology, she renders visible extraterrestrial perspectives of the heavens, which is to say, views of space from familiar planets such as Mercury or Venus, as well as from less familiar ones, such as Gliese 581c – a presumably uninhabited planet that lies 20 light years away in the Libra constellation. Bulloch plays here with modes of representation – and of illusion. For at first glance the viewer seems to recognize something familiar. Only upon closer analysis does it become apparent that the artist is presenting views unattainable for lack of adequate transport and sufficient time, and confronting us with wholly unfamiliar perspectives.

In summary, it can be said that "the" image of outer space does not exist, but rather that every historical period develops its own image. In the conception of distant worlds, as in other fields, popular and scientific styles of composition intersect and inspire each other. The experience of outer space, however, is a special case, with the popular imagination not only shaping this new and undiscovered world long before humankind could actually see or photograph it, but also influencing how the "real" images are processed.

8
The Very Idea of Non-Reductionist Science

Throughout the previous chapters we have talked about a non-reductionist science. In this chapter, we want to clarify what that means. We first look at the very successful notion of scientific reductionism as it gets used in the natural and social sciences. We then focus on some complications for the reductionist project in cognitive science that derive from embodied approaches to cognition, and ask we how it's possible to do science in this context.

1 Scientific reductionism

Science is typically considered to be reductionist, although there are also anti-reductionists among scientists and philosophers of science. What does it mean to be reductionist or anti-reductionist? One can typically find statements such as "heat reduces to kinetic molecular energy" which means (1) scientists can offer an explanation of heat in terms of kinetic molecular energy and/or (2) heat *just is* kinetic molecular energy. The first is a form of explanatory reduction; the second, ontological reduction. Explanatory reduction means, on a somewhat standard view, that one theory or vocabulary can be translated into another more basic theory or vocabulary, and that the latter explains more about the subject matter (see, e.g., Oppenheim and Putnam 1958). We know more about what heat is if we say 'kinetic molecular energy,' than if we just say 'heat.' There is more to say, of course, and someone might object that to know that heat is kinetic molecular energy doesn't tell us everything we need to know about heat – e.g., that it makes us sweat, that too much of it is not good for our health or for the earth, that too much of it may cause the icecaps to melt and lead to geopolitical crises, and so on. Supposedly, some of this could be explained in terms of kinetic molecular energy, but not all of it. One could argue that even if we have a reductionist story about heat in physics and chemistry, we would still need further explanations in biology and political science, and it's not at all clear that the concepts we would need in political science to explain the social effects of a local drought or a global crisis, for example, could be reduced to an explanation in terms of elementary particles.

On the one hand, it's important to note that some form of reductionism is not necessarily a bad thing in the realms of physics and chemistry. Some scientists and philosophers claim that it offers explanatory power and has driven the progress of science in these areas. On the other hand, it's not clear that it's a good thing in political science or the other social sciences. If one is tempted to say that reductionism is most appropriate in the natural sciences, then the issue is whether one can have a natural science of consciousness, cognition, the mind, self, free will, etc., if that means a reductionistic science.

In philosophy of mind and cognitive science, reductionist positions usually claim that mind is reducible to physical processes, and these physical processes are standardly understood to be neural processes, which in turn are reducible to molecular processes. Materialism or physicalism are standard in cognitive science, and these positions are usually thought to be reductionist so that neuronal (or, according to some, perhaps molecular or quantum) processes are thought to be the base level to which one reduces everything else. Looking in the opposite direction, one might think that idealists are inflationists rather than reductionists, but in fact anything like a claim that everything is mind (as one might find in the eighteenth century idealist, George Berkeley) is just another form of reductionism where the base is composed of mental events. Again, however, scientific materialism is generally the rule in science. Thus, Carnap states:

> ... science is a unity, [such] that all empirical statements can be expressed in a single language, all states of affairs are of one kind and are known by the same method. (Carnap 1934, p. 32)

Carnap here expresses ontological reductionism (all states of affairs are of one kind), and explanatory reductionism (all empirical statements can be expressed in a single language and known by the same method).

Eliminativism is a form of reductionism that denies the reality of the thing that is reduced. If the mind is reduced to the brain, then the mind is not real; only brain states are real. As van Riel and Gulick (2014, §1.) suggest, however, reductionists are generally realists about the reduced phenomenon: "They are, in that respect, *conservative*. They are committed to the reality of the reducing base and thus to the reality of whatever reduces to that base. If thoughts reduce to brain states and brain states are real, then so too are thoughts." It you are an eliminativist about the mind, however, and you think that only brain states exist, you find yourself in some very paradoxical positions. If only brain states exist, then phenomena like consciousness, self, free will, and perhaps even the world as we know it, don't exist – they are just illusions perpetrated by the brain. If one were to go that far, then indeed even what we call the brain, as part of the world as we know it, is an illusory construct of neural processes, which, in turn, are part of that thing that is only an illusory construct. In that case, it's not clear whether anything is real. *Au revoir* spouse, family, friends, home, job, you, me, and whatever. In this respect, eliminativist neuralism is just about as good as Berkeleyan idealism.

If, then, we stay with conservative reductionism, what do we get? The most reasonable way to think of it is to maintain that to explain something like experience we have at least two possible vocabularies – the mental and the physical. This is a position clearly expressed by Spinoza:

> The mind and the body are one and the same thing, which is conceived now under the attribute of thought, now under the attribute of extension.... Hence the order of actions and passions of our body is, by nature, at one with the order of actions and passions of the mind. (1970, III, Prop. 2s)

It's not clear that we avoid dualism, or at least property dualism, by preferring one vocabulary over the other – i.e., by reducing mind to brain. Even if one claims that a belief is really a set of neurons in the prefrontal cortex activated or maintained as a pattern for potential activation, one might be claiming that this set of neurons has dual properties – one property expressible in the vocabulary of the mental, and one property expressible in the vocabulary of the physical.

In contemporary discussions, whether we consider radical eliminitivism or conservative explanatory reductionism, the only elements that tend to be relevant in regard to understanding experience are either mental states (e.g., beliefs, thoughts, desires) or brain states or processes. Supposedly, anything else, such as some extra-neural bodily state, or some aspect of the environment, or some force of culture doesn't even enter into the explanation. All of these things have already been either reduced or eliminated, or in any case excluded from contributing to an explanation. Thus, an agent's body only enters into the explanation in terms of body (or B-formatted) representations in the brain (e.g., Gallese 2014; Goldman 2012). The cultural environment has an effect on cognition only if it is in some way represented in the brain.

This exclusion of factors that exist outside of the head – outside of mental and/or neural processes – is also a kind of reductionist move that one frequently finds in cognitive science. In some regards, this may be the result of a division of labor in science. Cognitive neuroscientists, for example, rightly focus their study on the brain – that's what they are supposed to explain. They are trying to work out how brain states cause mental states (or how cognition is reducible to neural states). They may not explicitly exclude other factors, such as bodily and environmental factors, or they may not deny, if asked, that extra-neural factors enter into the story in some way. But when they do their science, they control for all of those other things and end up explaining cognition totally in terms of neuronal processes. Sometimes neuroscientists, along with neuro-philosophers, end up making large but extraordinarily narrow claims about the nature of the mind. The mind just is a set of neural processes; the self is nothing more than a self-model generated by the brain (Metzinger 2004); *you* are nothing but a pack of neurons (Crick 1995), or, for all practical purposes, embodiment just is a set of B-formatted representations in the brain. The idea that *the body* as an actual physical organism, and *the environment* as a physical, social and cultural set of contexts, enter into and may have a causal or constitutive role to play in cognition is simply excluded, whether intentionally (as in the case of internalists – i.e., all relevant processes just are in the head), or unintentionally (because scientists are busy doing their own particular science and don't have time to think about these other things). This is a kind of exclusionary reductionism.[1]

For our purposes, in regard to claims about human experience, we want to steer away from exclusionary and other forms of reductionism. We want to defend a non-reductionist position that, more positively, is close to what Sandra Mitchell (2002) calls 'integrative pluralism'. We prefer the more positive formulation to distinguish it from classic discussions of non-reductionism. The latter focus on the idea that

[1] There may be interesting connections or disconnections between this notion of exclusionary reductionism and the more standard forms of reduction described above; this is an issue beyond the scope of this our analysis here. We thank Patrick McGivern for calling this to our attention.

sciences like psychology and economics, and so forth, offer explanations thought to be in some sense autonomous from underlying biological or physical explanations. The idea is always that psychological explanations are themselves complete and independent from neurological or physical ones. In the case of integrative pluralism, however, we have multi-scale explanations involving factors at various scales (neuroscientific, psychological, phenomenological, social, and so on) all contributing to an integrated explanation (McGivern 2008), in contrast to unconnected, multiple explanations that might run in parallel.

2 Integrated approaches to cognition

Some forms of phenomenology also involve explanatory reductionism. The notion of the "phenomenological reduction," which we've employed as part of the phenomenological method, is not equivalent to a materialist or neural reductionist strategy. As one finds it practiced in someone like Husserl (2012), however, it leads to a form of transcendental idealism in which we consider anything only from the perspective in which it appears to us in our experience. Methodologically, and for purposes of explaining how meaning is constituted, everything (including you and everything else in the world) is reduced to my experience of it. This is not an ontological claim, however; rather, it is a methodological reduction in the service of explanation. On the one hand, and in an odd way, at least in the way that some philosophers understand it, transcendental phenomenology might be thought to put us in the same neighborhood as naturalistic neural reductionism – that is, *in the head*, where, purportedly, everything of importance for cognition happens. On this view, phenomenologists are talking about the same thing as neuroscientists, but are simply adopting a vocabulary located on the other side of an unappreciated property dualism. On the other hand, many phenomenologists argue that the methodological (transcendental) reduction does the exact opposite; it puts us in the world, as we experience it, in a way that transcends the distinction between internal and external (Gallagher and Zahavi 2012).

Phenomenologists have developed a phenomenological psychology that still focuses on experience but also recognizes the role of the body and environment in constituting those experiences. The idea of a naturalized phenomenology (Petitot et al. 1999), which includes neurophenomenology (Varela 1996) and 'front-loaded' phenomenology (Gallagher 2003), although still philosophically controversial (see e.g.,

Zahavi 2010), takes phenomenological psychology into the lab and theoretically explores how embodied and enactive processes contribute (causally or constitutionally) to experience. A good model for this theoretical approach among the classic phenomenologists can be found in the work of Merleau-Ponty (2012; 1964). Although he did not conduct experiments, or participate in teams that carried out empirical studies, he paid close attention to the experimental literature and allowed it to inform his phenomenological writings.

Philosophical phenomenology and phenomenological psychology have provided important resources for the more recent development of embodied cognition and its variations: embedded (or situated), extended (or distributed), and enactive approaches to cognition. The '4Es,' as they are sometimes referred to (e.g., Rowlands 2010) – embodied, embedded, extended and enactive – really mark a sea change in cognitive science, away from internalist (purely 'in the head') approaches. The 4Es take into account bodily, affective, intersubjective, and environmental aspects of cognition, where environmental includes physical, social and cultural elements. We provide a brief summary here of the 4Es in order to indicate how they lead to an integrative cognitive science, but also to show (1) that the 4Es do not yet provide a unified account, and (2) that empirical science is still wrestling with questions about how to incorporate these approaches into their experimental paradigms.

2.1 Embodied cognition

The term ,'embodied,' in embodied cognition tends to be taken broadly to apply to a range of proposals (including enactive and extended approaches) that suggest that extra-neural bodily structures and processes have an effect on experience and cognition. For example, anatomy and bodily movement can be important, non-trivial factors in cognition. They can contribute to the shaping of cognition prior to brain processing (pre-processing) and subsequent to brain processing (post-processing) of information in the system defined to include brain, body, and environment (e.g., Chiel and Beer 1997; Gallagher 2005; Shapiro 2004; Straus 1966). For example, the fact that we have two eyes, positioned as they are, delivers binocular vision and allows us to see the relative depth of things. Similar things can be said about the position of our ears and our ability to tell the direction of sound. As Shapiro puts it, "the point is not simply [or trivially] that perceptual processes fit bodily structure. Perceptual processes *depend on and include* bodily structures" (2004, p. 190).

Our sensory experience also depends on the way our head and body move, as we see in the case of parallax (Churchland, Ramachandran, and Sejnowski 1994; Shapiro 2004) – an important principle that we had to take into consideration in designing the virtual aspects of our simulations. For example, stars in the background had to appear to move at a different rate than stars or planets in the foreground. Likewise, motor responses, rather than fully determined at brain-level, are mediated by the design of muscles and tendons, their degrees of flexibility, their geometric relationships to other muscles and joints, and their prior history of activation (Zajac 1993). Movement is not always centrally planned in brain processes; it is based on a competitive system that requires what Andy Clark terms 'soft assembly.' The nervous system learns to account for a variety of parameters, e.g., stiffness of limb or joint or level of muscle fatigue, which will then "*interact* with intrinsic bodily and environmental constraints so as to yield desired outcomes" (Clark 1997, p. 45).

Proprioceptive and affective (emotion-related) processes also have a profound effect on perception and thinking. For example, vibration-induced proprioceptive patterns that change the posture of the whole body are interpreted as changes in the perceived environment (Roll and Roll 1988, p. 162). Proprioceptive adjustments of the body schema can help to resolve perceptual conflicts (Harris 1965, p. 419; Rock and Harris 1967). Experimental alterations of the postural schema lead to alterations in space perception and perceptual shifts in external vertical and horizontal planes (Bauermeister 1964; Wapner and Werner 1965). Likewise, hormonal changes – changes in body chemistry – as well as visceral and musculoskeletal processes, can bias perception, memory, attention, and decision-making (Damasio 1994; Bechara et al. 1997; Gallagher 2005; Shapiro 2004). The regulation of body chemistry is not autonomous from cognitive processes, and vice versa. "Body regulation, survival, and mind are intimately interwoven," (Damasio 1994, p. 123).

One solid example of this is a study by Dansiger et al. (2011). They show that judges in the process of sentencing criminals are approximately 66% more lenient early in the morning than just prior to lunch, with a gradual decline throughout the morning; and then just after lunch they become 66% more lenient than before lunch. The judge's level of satiation has an effect on his judgment and decision process. As part of our protocol in screening participants for the experiments, we included questions about when they last ate and consumed alcohol. For example, we excluded participants who drank alcoholic beverages in the 24 hours prior to the experiment.

2.2 Embedded cognition and ecological factors

The physical environment plays a significant role in experience. Embodied approaches to cognition suggest that changes to the environment can affect cognition just as much as changes in the brain – and of course, the main idea is that changes in the brain are in fact correlated with changes in the environment. Consider, for example, cases of depression. A psychiatrist may claim that depression is due to specific aspects of brain chemistry – levels of serotonin or other neurotransmitters. This certainly may be true. One can also say, however, that depression may be due to current environmental factors – e.g., the absence of sunlight in the prolonged winters in northern climates. Therapies can be internalist (take these medications to balance your serotonin levels) or externalist (use these sun lamps or take a holiday in Spain). A change in environment may have the same effect as using a drug to induce changes in neurotransmitter levels. Even chronic clinical depression can be addressed by changes in life style and environments – including social environments.

More generally, cognition is always situated in physical, social, and cultural environments. The notion of ecological perception, as developed in the work of J. J. Gibson (1977; 2013) and his followers, builds on the fact that there is a tight fit between bodily movement and our experience of the environment. Any movement of our body will have an effect on our sensory experience of the world. This is most clear in the case of vision. Move your head from one side to the other, or up and down, and your spatial perspective on the world changes, even if only in minimal fashion. Such bodily movement, however, provides information to the visual system that allows it to maintain a coherent sense of the world as a set of locally stable structures. Changes in the environment register differently in the system despite the fact that they may produce identical effects in purely visual experience. Moving the whole visual environment two inches to the left may be visually equivalent to a slight movement of one's head to the right.

In fact, however, there is no purely visual experience. All vision is embodied in the sense that it is a product of a combination of bodily (e.g., motor and proprioceptive) and environmental factors. Our system can be rarely tricked, as in the example of having the experience that you and your train are moving, when it's really the train next to you pulling out of the station; or sitting in a plane as it is taking off and visually experiencing the interior of the plane as angled upward. The plane, of course, is objectively maintaining an upward angle, but since you are in the plane and also maintaining that same angle, you should not see a difference. Your proprioceptive system registers an upward angle causing a visual adjustment to create this effect.

In setting up our experiments, we also had to consider that in the human body visual, motor, vestibular, and proprioceptive systems are coupled. This was the previously mentioned worry about the effect that weightlessness might have on vision. We know, however, that after a few days, astronauts adjust to microgravity to the extent that it becomes less and less a topic of discussion. But we also know that the proprioceptive system recalibrates to microgravity after about 3 days (see NASA Science News 2001). Jonathan Cole, one of the neuroscience consultants on our project, conducted an informal experiment with astronauts on the space shuttle several years ago. Shortly after they attained orbit, he asked them to sit, close their eyes, and touch their knee. They couldn't find their knee on the first try. Their proprioceptive sense, normally attuned to the earth's gravitational field, was out of tune. But after a short amount of time, and certainly after 3 days, their proprioceptive systems adapt and all astronauts are able to find their knees. For this reason, we discounted any problems connected with the proprioception-vision link when, after a few days, astronauts were experiencing awe and wonder as they gazed out of the windows.

Ecological psychology also makes use of the concept of *affordance*. The notion of an affordance is an important one for understanding the way that body and environment are coupled. This is also an important concept for the notion of enactivism (see below). Something in the environment affords possibilities for action, but only relative to the perceiver's particular type of body and skill set. A chair affords sitting, but only for a body with bendable joints, and of a certain size. Most human adults easily perceive a chair as affording sitting; a lion does not. A swimming pool affords swimming, unless I don't know how to swim; in that case it may afford learning to swim, unless I am disabled in a way that would prevent that. A space shuttle affords flying, but only if I have the right kind of skills to fly it. A window affords visuals, but only if I have eyes to see and can gain a position to look out. Affordances, then, are not objective entities in the environment; nor are they just the skills I might have. Affordances are relational. They apply or exist only insofar as a body is embedded in an environment in an appropriate way.

Affordances can also be social. Gibson noted this in his original definition, and over the past ten years or so, this idea has been more fully explored. Other people afford certain possibilities that an individual agent may not have on his or her own. Moving a heavy box, for example, may not be possible alone. But if someone else comes into the room, and offers to help, moving the box suddenly becomes possible. Another person also affords communication, friendship, and many other things just by being another person.

Less developed in the literature is the notion of cultural affordances. If the other people are gathered in an institutional setting, following a set of rules that define how the institution works, for example, that situation itself affords the possibility of actions, for me or for them (and perhaps affords collective action on the part of the institution) that otherwise would not be possible. A space shuttle affords flying, but only if an institution like NASA, embedded in a national government, or a private corporation embedded in a vast industrial complex, has created a space shuttle and the tremendously complex infrastructure that supports such things. The possibility of flying on a space shuttle is a highly specialized cultural affordance.

2.3 Extended mind

The traditional cognitivist view of the mind is that it supervenes on brain processes. Neural processes are regarded as the "vehicles" or mechanisms that generate cognitive events. The idea of the extended mind is that entities in the environment can equally serve as vehicles of cognition. This is sometimes referred to as vehicle externalism. Clark and Chalmers (1998) explicate the idea with some thought experiments. If, for example, Otto, a person with Alzheimer's disease, relies on a notebook for recording useful information, and uses it, for instance, to find the location of a museum, then his cognition (his knowing where the museum is located) supervenes on the use of his notebook in a way that is similar to memory that supervenes on brain areas in someone who simply consults biological memory. The act of using the notebook is as much a cognitive act as the act of recollection. Thus, Clark (2008a) suggests that the extended cognitive system starts with the brain and includes body and environment. As he puts it, "the larger systemic wholes, incorporating brains, bodies, the motion of sense organs, and (under some conditions) the information-bearing states of non-biological props and aids, may sometimes constitute the *mechanistic supervenience base* for mental states and processes" (2008a, p. 38). This is an idea that goes back to the pragmatists. Thus, John Dewey once suggested:

> Hands and feet, apparatus and appliances of all kinds are as much a part of [thinking] as changes in the brain. Since these physical operations (including the cerebral events) and equipments are a part of thinking, thinking is mental, not because of a peculiar stuff which

> enters into it or of peculiar non-natural activities which constitute it, but because of what physical acts and appliances do: the distinctive purpose for which they are employed and the distinctive results which they accomplish. (Dewey 1916, pp. 8–9)

According to this view, cognition extends to the use of our bodies (e.g., our fingers for counting; our gestures for thinking), and to the use of tools and technologies as long as those uses contribute to the accomplishment of a cognitive task – remembering something, or solving a problem, for instance. This idea also extends to social relations and arrangements. For example, two people working together, relying on each other's resources, can remember more than the aggregate of what each can remember on his or her own (Sutton et al. 2010). Two people working together may be able to solve a problem that the two of them working independently could not solve. The notion of the socially extended mind (Gallagher 2013) suggests that individuals are able to extend their cognitive performance by engaging with social institutions, such as using the legal system to solve a problem. Indeed, science, as a social institution that involves certain practices and physical apparatus, is a cognitive enterprise that involves individuals in cognitive practices that extend beyond any one individual.

2.4 Enactive cognition

Enactive views on embodied cognition emphasize the idea that perception is *for action*, and that this action-orientation shapes most cognitive processes. This approach often comes with strong calls to radically change our ways of thinking about the mind and doing cognitive science (e.g., Gallagher and Varela 2003; Thompson 2007; Varela, Thompson and Rosch 1991). Following Clark (1999), Thompson and Varela (2001) offer a three-point summary of the enactive view:

1. understanding the complex interplay of brain, body and world requires the tools and methods of nonlinear dynamical systems theory;
2. traditional notions of representation and computation are inadequate;
3. traditional decompositions of the cognitive system into inner functional subsystems or modules ('boxology') are misleading, and blind us to arguably better decompositions into dynamical systems that cut across the brain–body–world divisions. (Thompson and Varela 2001, p. 418; also see Chemero 2009, p. 29).

Similar to the idea of extended cognition, enactive approaches argue that cognition is not entirely "in the head," but distributed across brain, body, and environment. More so than extended mind theorists, enactivists claim that (human) bodily processes shape and contribute to the constitution of consciousness and cognition in an irreducible and irreplaceable way. Specifically, on the enactive view, biological aspects of bodily life, including organismic and emotion regulation of the entire body, have a permeating effect on cognition, as do processes of sensorimotor coupling between organism and environment. Noë (2004; also see O'Regan and Noë 2001; Hurley 1998) developed a detailed account of enactive perception where sensory-motor contingencies and environmental affordances take over the work that classic cognitive science had attributed to neural computations and mental representations.

Intersubjective interaction is also an important dimension that enactivists consider. The environment, which helps to form our cognitive systems, includes the presence of other people, with whom we begin to interact from the very beginning. Developmental studies show that infants engage in embodied intersubjective practices from birth and develop through primary and secondary intersubjective processes (Reddy 2008; Trevarthen 1979). Primary intersubjectivity includes bodily interactions that exploit facial expression, posture, movement, gestures, and distinct forms of sensory-motor couplings. On the enactivist view, mirror neurons contribute to the direct social perception of motor intentions and to response preparation (Gallagher 2007). Context and social environment also contribute to "secondary intersubjective" practices starting at 9–12 months of age (Trevarthen and Hubley 1978). In the intersubjective context, perception is often *for inter-action* with others, where perceptually-guided interaction becomes a principle of social cognition and generates meaning in processes of 'participatory sense-making' (De Jaegher and Di Paulo 2007; De Jaegher, Di Paulo and Gallagher 2010; Gallagher 2009).

3 Doing science in a 4E context

On the one hand, these embodied, embedded, extended and enactive approaches share certain fundamental concerns and some common assumptions that contrast with classic computational, internalist, and reductionist models of the mind. On the other hand, between, and even within these different approaches, one can find ongoing debates over the precise roles of body and environment, the precise nature of the coupling between these elements, the role of representations in cognition, and philosophical debates about the nature of functionalism. To the extent that they can be resolved, the promise is of a more unified embodied theory of cognition. We need not enter into these differences and debates here, however (see Gallagher, in press). Yet, there is one issue that seems directly relevant to our study of awe and wonder. This concerns the question about the specific nature of consciousness and whether it, like other aspects of cognition (e.g., perception, memory, intentions, beliefs, etc.) is extended (whether it supervenes on environmental elements, or requires a certain kind of environmental coupling), or whether it depends exclusively on neural processes. To the extent that our study is about experience, and specifically about the forms of consciousness called awe and wonder, this issue seems relevant.

Andy Clark (2008b) for example, has argued that all forms of cognition except consciousness (i.e., phenomenal experience) can be analyzed in terms of the extended mind hypothesis. In more radical versions of enactivism, it's claimed that consciousness is also extended; that is, the physical basis of phenomenal experience is not confined exclusively to the brain, but extends beyond it by means of sensorimotor dynamics (e.g., Pepper 2014; Thompson and Stapleton 2009).

This issue involves a philosophical distinction between what is *constitutive* for something and what is merely *causal*. If something is constitutive of X, then it is a necessary part of X; if something is causal, it doesn't end up as a necessary part of X, even if it may play a necessary role in causing X. Consider, for example, making a cake. We would say that the flour, sugar and other ingredients that go into the cake are constitutive – they end up being part of the cake. There are also causal factors that are tied to making the cake – pouring and mixing things together. In some sense there would be no cake unless these causal factors do what they do; yet we wouldn't say that the cake is constituted by these causal factors.

To say what factors constitute consciousness, and what factors might be merely causal for consciousness is a much more difficult, and indeed, metaphysical issue. It brings us back to problems that involve the explanatory gap, and the "hard problem" of consciousness (see Chapter 4). Are neural processes actually constitutive of consciousness (as some internalists might argue), or are they merely a causal contributory? Does the phenomenality (the "what it is like") of experience depend constitutively on objects in the environment, where e.g., the experiential qualia of redness is nothing other than the redness of the apple that we perceive, or are those objects mere causal occasions for something that is intrinsically dependent on neural processes? There are genuine philosophical conundrums in these questions. Luckily we do not have to answer these questions or stake out a metaphysical position in order to perform the science. Science is interested in both causality and constitution, although it takes a huge amount of science to get into a position to say anything about the constitutive nature of reality (e.g., water just is H_2O). There is no such simple consensus statement about consciousness. There is no general agreement, for example, that consciousness is just the activation of some set of neural networks, or that it is its own unique element that is not reducible to neural activations, although there are philosophers and scientists who may defend such positions.

Again, however, we do not have to answer these questions in order to perform the science. Phenomenology, for example, suggests that we can simply bracket these metaphysical issues – i.e., we don't have to answer them in order to provide an account of the phenomena; and science can pragmatically agree with this. At the same time, by doing the phenomenology and doing the science we may get closer to answering (or reframing) such questions. In this respect, they remain relevant questions.

Even if we are not required to solve the hard problem of consciousness, we are still faced with the difficulties that embodied approaches to cognition pose for science. We discussed these issues in Chapter 4. Embodied cognition generally, in its more holistic formulations, contends that a scientifically adequate account of experience requires that we understand not just brain function, but also bodily responses in a physical, social and cultural environment. It thereby aspires to a non-exclusionary, integrative cognitive science. The broad focus requires that we gather a fuller set of data by including, not just EEG and fNIR recordings of brain function, but also aspects of bodily and affective (e.g., ECG) processes, the strictly physical (or in this case, virtual) environment, and selective information about the participants' cultural backgrounds. The broader vision for cognitive science includes the integration of a diverse set of factors that may not be reducible to exclusively neural frameworks.

These concerns guided the designs of both experiments. By using a neurophenomenological approach, our experiments resisted third-person objectification of phenomena that are rightly first-person subjective experiences. As we indicated in Chapter 4; however, the hard problem for us was the hard problem of doing science in a way that encompasses the holistic scope of embodied cognition.

If there is a clear philosophical tension between the strict, but productive limits that define scientific method and the philosophical holism that comes along with embodied approaches in their attempts to work out explanations in terms of the dynamics of brain-body-environment, this also turns into a pragmatic tension when one goes to do the science. The issue is how to combine established methods in psychology and the cognitive sciences that provide rigorous and respected procedures for experimentation, with the philosophical motivation of embodied/enactive neurophenomenology.

If embodied cognition, and especially the phenomenologically-inspired enactivist approach to embodied cognition, takes experience to be not just a matter of neuronal processing in the brain, but something that involves bodily (motor, affective, autonomic, and peripheral) processes that enact meaning as they are coupled to specific physical, social, and cultural environments, including the organism's (the subject's) previous history – that's quite a bit to fit into any one experiment. One response to that is to make a series of simple experiments that address each issue, and to make many such experiments that will eventually lead to a larger picture of the phenomenon at stake. This idea is not without merit and, indeed, is the usual road experimental psychology travels. On this road one can run into what we called the 'clunky robot problem' (in Chapter 3), however. That is, by studying the pieces, we don't necessarily learn how the dynamical relations between them work. Alternatively, one can try to build as much as possible into an experimental paradigm; but this is where philosophical holism comes into conflict with scientific control. Thus, from Experiment 1 we learned not only some things about awe and wonder, but some important lessons about trying to use this kind of holistic approach. This led to Experiment 2 where we negotiated on some of the elements, introduced more controls, but still retained a sufficient number of elements to satisfy our non-reductionist, integrationist requirements.

These negotiations between enactivist holism and the necessary requirements of scientific rigor reflect more generally the challenge presented by embodied cognition and the quest for a non-reductionist science. Embodied/enactivist approaches to cognition, by incorporating not just the brain, not just the environment, not just behavior, but the rich dynamics of brain-body-environment, insist on a holistic conception of cognition that is difficult to operationalize. But, as we hope we have shown, not impossible, despite the requirements of good experimental controls and the normal division of labor in science.

If both experimental design and data analysis is more difficult, we suggested, in the end it may be easier to include crucial factors than to ignore them. Including first-person reports (in the form of phenomenological interviews) not only seems essential to these kinds of experiments, it also provided important information from which we could draw in forming our interpretation of the other data. Maintaining precise control over the environment and simulation timeline allowed for a more precise mapping of descriptions of experience from the phenomenological interviews. The role of cultural background forces in shaping experience was shown to be richer and more complex than any one questionnaire could formulate. But this too was expanded by what we learned from the phenomenological interviews. All of these things helped us contextualize the physiological and neurophysiological data and allowed for a more coherent account of awe and wonder than we find in prior literature.

4 The sky is not the limit: some answers and more questions

Following the philosophical direction explicated in the previous sections, we attempted to integrate into our study of awe and wonder (and this generalizes to many types of consciousness) the idea that a full account will not be had simply by looking in the brain – although obviously the neuroscience is an important component. We attempted to understand precisely what elements in the environment are involved, what aspects of embodiment are involved, what aspects of culture and background practices are involved. The task was to find an experimental paradigm that could take all of these factors into account. We think the neurophenomenological approach goes some distance in this direction, and that it is easily supplemented using other instruments such as the ESSE and various questionnaires. Much of what we learned in this project concerns methodology. But, of course, we also learned something about the experiences of awe and wonder.

So what did we learn about awe and wonder? We can briefly summarize with the following oversimplified list. Awe and wonder (and related phenomena):

1. can be defined by more precise categories of experiences
2. may also escape all categories
3. can be replicated in simulated environments
4. are tied to visual experience (in the cases we examined)
5. are not necessarily tied to religiosity
6. may be shaped by culturally-related expectations
7. may be enhanced in people with certain self-identified profiles

8. involves complex patterns of lower alpha and increased beta neural oscillations

More details, however, are found in the experiments.

4.1 Experiment 1

The objective of Experiment 1 was to determine

1. The physiological markers associated with the experiences of awe and wonder (AW)
2. Which variables will differentiate between AW experiencers and AW non-experiencers identified from the interviews, and
3. The environmental factors that influence AW experiences.

Our results showed that several EEG metrics were able to differentiate between AW experiencers, and AW non-experiencers. Specifically, during the Earth condition increased (compared to baseline) left hemisphere (LH) and right hemisphere (RH) theta was found among non-experiencers of awe compared to experiencers of awe. Increased theta in RH and LH indicates increased drowsiness and fatigue (Paus and Zatorre 1997), a decreased awareness and ability to actively interact with the environment (Schacter 1977), and a non-attention to stimuli (Shiota et al. 2007) in AW non-experiencers. Clearly there is no marker of awe experience in this data, but it does confirm that the experience of awe requires attention to stimuli.

This interpretation was supported by the interviews, with the non-experiencers reporting boredom and inattention to the stimuli. During the interviews, AW experiencers also mention feelings of boredom and inattention but, on average, not until much later in the simulation; these feelings are likely associated with the theta changes occurring in the physiological measures around the 8–10 min period that resemble changes that occur during a vigilance task, reflecting difficulty in maintaining attention to the task over time, or increased workload (Reinerman-Jones et al. 2010).

Results also showed significant differences between AW experiencers and non-experiencers on two subscales of the BMMRS (Masters et al. 2009): *Experiential Comforting Faith* (ECF) and *Private Religious Practices* (PRP). These differences were able to differentiate traits between awe experiencers and non-experiencers during the Earth view. Awe non-experiencers were found to have higher levels of religiosity/spirituality compared to awe experiencers. Although awe experiences have been linked to religiosity (Emmons 2005; Newberg and Newberg 2005), the present results show that space-related awe experiences can occur without religiosity; this is consistent with Keltner and Haidt's (2003) model where perceived vastness and accommodation associated with AW do not depend on a person's religiosity.

According to the phenomenological interviews, the Earth condition elicited higher levels of AW compared to the Deep Space condition. These results indicate that participants viewed the Earth scenario as more powerful and moving than the Deep Space scenario. Participants had greater difficulty accommodating the Earth view into their current cognitive structures compared to the Deep Space view, again consistent with Keltner and Haidt's (2003) model of awe and wonder experience. These findings are further supported by the differences in EEG beta levels showing greater frontal lobe beta, parietal/occipital lobe beta, left hemisphere beta, and right hemisphere beta during the Earth condition compared to the Deep Space condition. Increases in beta have been linked to increases in arousal and attention (Prinzel et al. 2000), which indicates that participants were more aroused and attentive during the Earth condition compared to the Deep Space condition. Combined, these findings suggest that information-rich and attention-grabbing environments are influential in generating AW experiences, consistent with findings from Shiota et al. (2007).

4.2 Experiment 2

The second experiment was more tightly controlled with an emphasis on the role that visual stimuli play in AW experience. The visual simulations involved differences in starting location in the initial minute. FOC: focused on an aerial view of a familiar environment (the UCF campus); GLO: started with a global and unfamiliar view. Despite these context differences, there was no statistically significant difference between groups on their ESSE experiential indications. Participants in both conditions reported experiencing AW at later points in the simulation.

The EEG results, however, indicated a difference between the participants' experiences of FOC and GLO conditions. There was a drop in alpha in both groups, but with a greater drop in the FOC condition. As we saw, however, the changes in alpha were subject to several different interpretations, ranging from the effects of negative stimuli, to changes in lateral gaze, to shifts of attention or vigilance. In this respect, however, we were able to get a better picture of the significance of changes in alpha by comparing AW experiencers to non-experiencers. Across LH, occipital/parietal areas, and frontal lobe, AW experiencers showed greater suppression (below the mean difference from baseline) of alpha than the AW non-experiencers. Changes in alpha thus correlated significantly with respect to AW experiences.

The phenomenological interviews indicated that many AW non-experiencers felt relaxed, and even sleepy by the end of the experiment, correlating to increased theta. We suggested that future research should disentangle the relations between thoughtful relaxation or meditation and drowsiness which we also found reflected in the first-person articulation of experience in AW non-experiencers.

The ESSE explicitly asked participants to report aspects of AW; these reports were then correlated with both the results of the phenomenological interviews and the neurophysiological analysis. We found opposing correlative directions for visual processing-associated beta and theta behaviors in those who self-identified as "logical" rather than "spiritual" or "religious." This suggested that the spiritually inclined person not only sees the world differently in a figurative sense, but quite literally. A person who considers herself "spiritual" or "religious" will process the visual world differently on the neurophysiological level. Since neurological processes associated with experiences of awe and wonder are only *partially* the result of the presented context of the immediate stimulus, and since one's self-identification is closely tied to one's historical circumstance, personal experience, episodic memories, learned conceptual schemas, and cultural practices, we attempted to explore some of these factors through the use of questionnaires. Here we learned from the BMMRS that awe experiencers engaged in a lower level of religious practice than awe non-experiencers, and from the TAS questionnaire that awe experiencers were less likely to become absorbed in sensory experience. To gain a full understanding of these suggestive discoveries requires further research.

4.3 Further questions

Like most scientific projects, our project produced more questions than answers. For example, we suggested that the issues we just mentioned can raise a hermeneutical question about the nature of the phenomenological interview itself. If a speaker (participant) gives an account of experience from his or her personal perspective, the listener (interviewer), in a way, is exposed to a worldview that is, in experiential terms, more or less "spiritual" or "logical," etc. – a worldview connected to the speaker's self-identification. Self-identification (as spiritual or logical, etc.) may act as a type of perceptual filter or frame, not always recognized by others or even by oneself. In the interview process, as in our interactions with others generally, differences in such frames may create roadblocks to understanding, just as recognizing such differences may facilitate communication.

The self-described "logical" person may not only see something different from the self-identified "spiritual" person, but also in conversation, may draw the listener into a world that is shaped by that experience. The structure of the self-identification/perceptual filter can shape the second-person interaction within the interview process and influence our understanding of the experience. In this regard, however, the ESSE and other questionnaires helped us understand the particular dynamics of the interview process. Neurophenomenology, supplemented with these additional measures, can contribute to a ground-level mapping of these difficult and entangled aspects of experience and communication in an interactive world.

It is certainly an interesting question to consider whether the way that someone self-identifies (as logical, spiritual or religious) can affect their perceptual experiences, or, beyond that, whether socio-political views may be closely bound to different perceptual experiences. These are large questions. There are, however, two more protracted questions that our current approach may be able to answer, and that we plan to explore in future experiments. First, to what extent would occurrent social interaction modulate experiences of awe and wonder? Would being with others in the simulated environment lead to more or less, or different experiences of awe and/or wonder? Second, are there any long-term effects that result from these experiences? Some of the astronauts who have had such experiences have, after returning to earth, pursued lives that were reoriented, or more fully oriented towards ethical and spiritual, and sometimes religious interests. Could the kind of brief experiences of awe and wonder generated in simulated environments have similar effects?

We think that the experiments summarized here are good examples of how to do a multi-dimensional study that integrates first-person (phenomenological) data with third-person measures (including neuro- and physio data, as well as psychometric data and data on the cultural practices of the participants), thereby providing a fuller picture of the subject matter. This large methodology, which combines information from phenomenology, psychology, and neuroscience in a tightly controlled environmental setting, we propose, is one way to do a non-reductionist, integrative-pluralist cognitive science and to study consciousness more generally. Since cognition is embodied and situated in rich social and cultural environments, not all causal, or constitutive factors of experience are to be found simply in the brain. An integrative cognitive science attempts to grasp as many of these non-neural factors as possible, without ignoring the important role of brain processes. Even to understand what the brain is doing, however, we need the broader picture that involves experiential, embodied, socially and culturally situated factors that contribute to make each person's experience what it is.

Appendix: The Experiment-Specific Survey of Experience (ESSE)

Experiment-Specific Survey of Experience

Please answer each of the questions to your best ability.

STOP! The research assistant must verify that your PARTICIPANT IDENTIFICATION NUMBER is entered correctly.

[ ]

Demographics Questionnaire for *Viewing Earth from Space: First-Person Experiences*
What is your sex?

- ○ Male
- ○ Female

What is your age?

[ ]

What is the HIGHEST level of education you have COMPLETED?

- ○ High School
- ○ Associates Degree or 2 years of College/University
- ○ Bachelors Degree
- ○ Masters Degree
- ○ Doctoral Degree

When did you use computers in your education? *Select all that apply.*

- ☐ Preschool
- ☐ Grade School
- ☐ Junior High/ Middle School
- ☐ High School
- ☐ Technical School
- ☐ College
- ☐ Did not use

What is your major?

[ ]

What is your minor? Please enter "NA" if you don't have one.

[ ]

Are you in your usual state of physical health?

- ○ Yes
- ○ No (please explain) [ ]

Where do you currently use a computer? *Select all that apply.*

- ☐ Home
- ☐ Work
- ☐ Library
- ☐ Other (specify) [ ]
- ☐ Do not use

Is English your native (first) language?

- ○ Yes
- ○ No

At what age did you begin speaking English?

[ ]

Would you consider yourself a fluent speaker of English?

- ○ Yes
- ○ No

Would you consider yourself a fluent reader of English?

- ○ Yes
- ○ No

Do you typically understand spoken English without difficulty?

- ○ Yes
- ○ No

What was your first language?

How many hours per day do you spend WORKING on a computer?

- ○ 0
- ○ <1
- ○ 1–2
- ○ 3–4
- ○ 4–5
- ○ 5–6
- ○ 7+

How many hours per day do you spend READING?

- ○ 0
- ○ <1
- ○ 1–2
- ○ 3–4
- ○ 5–6
- ○ 7+

How many hours per day do you spend WATCHING TV?

- ○ 0
- ○ <1
- ○ 1–2
- ○ 3–4
- ○ 5–6
- ○ 7+

Approximately how many hours of sleep did you get last night?

Which of the following best describes your expertise with computers?

- ○ Novice
- ○ Good with one type of software package (such as word processing or slide shows)
- ○ Good with several software packages
- ○ Can program in one language and use several software packages
- ○ Can program in several languages and use several software packages

How often do you ...	Never	Less than Once a Month	Once a Month	2–3 Times a Month	Once a Week	2–3 Times a Week	Daily
Use graphics or drawing features in software packages?	○	○	○	○	○	○	○
Go to movies?	○	○	○	○	○	○	○
Watch IMAX or surround-screen movies?	○	○	○	○	○	○	○
Go to theme parks/ amusement parks?	○	○	○	○	○	○	○
Play video/computer games?	○	○	○	○	○	○	○
Visit a museum?	○	○	○	○	○	○	○
Visit a planetarium?	○	○	○	○	○	○	○
Attend faith-based or religious activities?	○	○	○	○	○	○	○

Which types of computer/video games do you most often play? *Select all that apply.*

- ☐ Action (First person shooter, fighting, etc.)
- ☐ Adventure, Real-time 3D
- ☐ Role Playing (including MMOs)
- ☐ Simulation (Sims, Civilization, etc.)
- ☐ Strategy/Puzzle
- ☐ Party, dance, or music
- ☐ Sports
- ☐ Other

When you do PLAY VIDEO GAMES, how many hours per day do you spend?

- ○ 0
- ○ <1
- ○ 1–2
- ○ 3–4
- ○ 5–6
- ○ 7+

Which of the following amusement/entertainment sites have you visited?

- ☐ Disney parks (i.e. Disneyworld, Disneyland, Euro-Disney)
- ☐ Disney Quest
- ☐ Universal Studios/ Islands of Adventure
- ☐ Kennedy Space Center
- ☐ Busch Gardens
- ☐ Sea World
- ☐ Six Flags

Rank your own level of competency with graphics or drawing software.

	Minimal skill or experience	Moderate skill level	Advanced or professional level of skill in at least one graphic or drawing software
My level of experience/ competency with graphic software is ...	○	○	○

STOP!

You have completed the demographic portion of this questionnaire. Wait for the research assistant to give you further instructions. Research Assistant Code

☐

The Research Assistant will read the following aloud. Wait until it is read before continuing:

The following questions will help us interpret the results from your interview and physical readings more accurately. We will be looking especially at indicators of emotional experiences. To help you describe your experience, we ask that you make the following distinctions:

When we use the word AWE, we mean: *a direct and initial feeling when faced with something incomprehensible or sublime.*

When we use the word WONDER, we mean: *a more reflective feeling one has when unable to put things back into a familiar conceptual framework.*

When we use the word CURIOSITY, we mean: *wanting to know*, see, experience, understand more. When we use the word HUMILITY, we mean: a sensation about *one's relation to the universe or one's significance*.

While there may be other ways to use these terms, these are the definitions we are using in the following questions.

Use the sliding scale to show the degree to which each statement describes you.

	This does not describe me at all									This is a perfect description
	0	10	20	30	40	50	60	70	80	90 100
I am a spiritual person										
I am a logical person										
I am a religion person										
When viewing the images in today's experiment, I experienced AWE.										
When viewing the images in today's experiment, I experienced WONDER.										
When viewing the images in today's experiment, I experienced CURIOSITY.										
When viewing the images in today's experiment, I experienced HUMILITY.										
Soemthing about this experiment felt familiar to me.										
I am generally a reflective person										

Which best describes your experience? I experienced AWE the MOST when viewing...

- ○ close images of the Earth (toward the beginning of the video).
- ○ distant images of the Earth (toward the end of the video).
- ○ the images of the geometric shape.

Which best describes your experience? I experienced WONDER the MOST when viewing...

- ○ close images of the Earth (toward the beginning of the video).
- ○ distant images of the Earth (toward the end of the video).
- ○ the images of the geometric shape.

Which best describes your experience? I experienced CURIOSITY the MOST when viewing...

- ○ close images of the Earth (toward the beginning of the video).
- ○ distant images of the Earth (toward the end of the video).
- ○ the images of the geometric shape.

Which best describes your experience? I experienced HUMILITY the MOST when viewing...

- ○ close images of the Earth (toward the beginning of the video).
- ○ distant images of the Earth (toward the end of the video).
- ○ the images of the geometric shape.

References

Baigrie, B. S. (ed.) 1996. *Picturing Knowledge. Historical and Philosophical Problems Concerning the Use of Art in Science*. Toronto: University of Toronto Press.

Barber, S. 2012. *Muybridge: The Eye in Motion*. Washington, DC: Solar.

Bauermeister, M. 1964. The effect of body tilt on apparent verticality, apparent body position and their relation. *Journal of Experimental Psychology* 67, 142–147.

Bayne, T. 2004. Closing the gap? Some questions for neurophenomenology. *Phenomenology and the Cognitive Sciences* 3(4), 349–364.

Bechara, A., Damasio, H., Tranel, D. and Damasio, A. R. 1997. Deciding advantageously before knowing the advantageous strategy. *Science* 275 (5304), 1293–1295.

Bickle, J. 2003. *Philosophy and Neuroscience: A Ruthlessly Reductive Account*. Boston: Kluwer Academic.

Biddle, W. 2009. *Dark Side of the Moon: Wernher von Braun, the Third Reich, and the Space Race*. New York: WW Norton & Company.

Bockelman, P. S. 2013. *Neurophenomenological Methods: Experiences of Earth and Space in Simulation* (Unpublished doctoral dissertation). University of Central Florida, Orlando, Florida.

Bockelman, P., Reinerman-Jones, L. and Gallagher, S. 2013. Methodological lessons in neurophenomenology: review of a baseline study and recommendations for research approaches. *Frontiers in Human Neuroscience*, 7, 608. doi:10.3389/fnhum.2013.00608

Bonnefond, M. and Jensen, O. 2012. Alpha oscillations serve to protect working memory maintenance against anticipated distracters. *Current Biology* 22(20), 1969–1974. doi:10.1016/j.cub.2012.08.029

Bonner, E. T. and Friedman, H. L. 2011. A conceptual clarification of the experience of awe: an interpretative phenomenological analysis. *Humanistic Psychologist*, 39(3), 222–235.

Bono, M. 2011. *Thomas Ruff: Stellar Landscapes*. Heidelberg: Kehrer.

Brand, S. 2009. Photography changes our relationship to our planet. Smithsonian Photography Initiative. Cited at http://en.wikipedia.org/wiki/Stewart_Brand.

Bredekamp, H. 2011. Blue marble: der blaue Planet. In C. Markschies, I. Reichle, J. Brüning and P. Deuflhard (eds) *Atlas der Weltbilder* (366–375). Berlin: Akademie-Verlag .

Bulkeley, K. 2002. The evolution of wonder: religious and neuroscientific perspectives. Paper presented at the annual meeting of the American Academy of Religion, Toronto, Canada (November 2002).

Byers, B. K. 1977. *Destination Moon: A History of the Lunar Orbiter Program*. NASA Technical Memorandum X-3487. Washington, DC: NASA History Office.

Cacioppo, J. T. and Petty, R. E. 1982. The need for cognition. *Journal of Personality and Social Psychology* 42(1), 116–131.

Cahn, B. and Polich, J. 2006. Meditation states and traits: EEG, ERP, and neuroimaging studies. *Psychological Bulletin* 132(2), 180–211.

Cannon-Bowers J. A., Salas E. and Converse S. 1993. Shared mental models in expert team decision making. In N. J. Castellan Jr. (ed.) *Individual and Group Decision Making: Current Issues* (221–246). Hillsdale, NJ: Lawrence Erlbaum Associates.

Carnap, R. 1934. *The Unity of Science*, London: Kegan Paul, Trench, Trubner, and Co.

Carpenter, S. 1962. *We Seven: By the Astronauts Themselves*. New York: Simon and Schuster.

Chalmers D. J. 1995. Facing up to the problem of consciousness. *Journal of Consciousness Studies* 2(3), 200–219.

Chemero, A. 2009. *Radical Embodied Cognitive Science*. Cambridge, MA: MIT Press.

Cheron, G., Leroy, A., Palmero-Soler, E., De Saedeleer, C., Bengoetxea, A., Cebolla, A. M., Vidal, M., Dan, B., Berthoz, A. and McIntyre, J. 2014. Gravity influences top-down signals in visual processing. *PloS One* 9(1), e82371.

Chiel, H. and Beer, R. 1997. The brain has a body: adaptive behavior emerges from interactions of nervous system, body and environment. *Trends in Neuroscience* 20, 553–557.

Chudleigh, M. 1703. *Poems on Several Occasions. Together with the Song of the Three Children Paraphras'd*. London: Bernard Lintott at the Middle Temple Gate in Fleetstreet. pp. 161–162.

Churchland, P. S. 2011. *Braintrust: What Neuroscience Tells Us about Morality*. Princeton: Princeton University Press.

Churchland, P. S. 2013. *Touching a Nerve: Our Brains, Our Selves*. New York: WW Norton & Company.

Churchland, P. S. Ramachandran, V. S. and Sejnowski, T. J. 1994. A critique of pure vision. In C. Koch and J. L. Davis (eds) *Large-scale Neuronal Theories of the Brain*. Cambridge, MA: MIT Press.

Clark, A. 1997. *Being There*. Cambridge, MA: MIT Press

Clark, A. 1999. An embodied cognitive science? *Trends in Cognitive Sciences* 3(9), 345–351.

Clark, A. 2008a. *Supersizing the Mind: Reflections on Embodiment, Action, and Cognitive Extension*. Oxford: Oxford University Press.

Clark, A. 2008b. Pressing the flesh: a tension on the study of the embodied, embedded mind. *Philosophy and Phenomenological Research* 76, 37–59.

Clark, A. and Chalmers, D. 1998. The extended mind. *Analysis* 56(1), 7–19.

Clarke, A. C. 1973. *Profiles of the Future: An Enquiry into the Limits of the Possible*. 2nd Ed. London: Phoenix.

Clément, G. 1998. Alteration of eye movements and motion perception in microgravity. *Brain Research Reviews* 28(1), 161–172.

Clément, G., Lathan, C., and Lockerd, A. 2008. Perception of depth in microgravity during parabolic flight. *Acta Astronautica* 63(7), 828–832.

Clément, G. and Reschke, M. F. 2008. *Neuroscience in Space*. New York: Springer.

Coltheart, M. 1981. The MRC psycholinguistic database quarterly. *Journal of Experimental Psychology* 33A, 497–505.

Crick, F. 1995. *The Astonishing Hypothesis: The Scientific Search for the Soul*. New York: Simon and Schuster.

Damasio, A. 1994. *Descartes Error: Emotion, Reason, and the Human Brain*. New York: G. P. Putnam.

Danziger, S., Levav, J. and Avnaim-Pesso, L. 2011. Extraneous factors in judicial decisions. *Proceedings of the National Academy of Sciences* 108(17), 6889–6892.

Daston, L. and Galison, P. 2007. *Objectivity*. New York: Zone Books.

De Jaegher, H. and Di Paolo, E. 2007. Participatory sense-making: an enactive approach to social cognition. *Phenomenology and the Cognitive Sciences* 6, 485–507.

De Jaegher, H., Di Paulo, E. and Gallagher, S. 2010. Can social interaction constitute social cognition? *Trends in Cognitive Sciences* 14(10), 441–447

Dennett, D. C. 1989. *The Intentional Stance*. Cambridge, MA: MIT press.

Dennett, D. C. 1991. *Consciousness Explained*. Boston: Little, Brown, and Co.

De Toffol, B., Autret, A., Degiovanni, E. and Roux, S. 1990. Spectral analysis of the EEG (alpha rhythm) and activity in the left hemisphere: the effects of lateral gaze. *Neuropsychologia*, 28(8), 879–882.

Dewey, J. 1916. *Democracy and Education*. Carbondale: Southern Illinois University Press, 1985.

Didonna, F. (ed.) 2009. *Clinical Handbook of Mindfulness*. Berlin: Springer.

Dirican, A. C., and Göktürk, M. 2011. Psychophysiological measures of human cognitive states applied in human computer interaction. *Procedia Computer Science* 3, 1361–1367.

Dominey, P. F., Prescott, T., Bohg, J., Engel, A. K., Gallagher, S. Heed, T., Hoffmann, M., Knoblich, G., Prinz, W. and Schwartz, A. (in press). Implications of action-oriented paradigm shifts in cognitive science. In A. K. Engel, K. Friston, and D. Kragic (eds), *Where's the Action? The Pragmatic Turn in Cognitive Science*. Cambridge, MA: MIT Press.

Douy, M. and Douy, J. 1993. *Décors de Cinéma: Les Studios Français de Méliès À Nos Jours*. Paris: Éd. du Collectionneur.

Dukes, S. 1984. Phenomenological methodology in the human sciences. *Journal of Religion and Health* 23(3), 197–203.

Eisler, C. 2002. La 'Tempesta' di Giorgione: Il primo 'Capriccio' della pittura Veneziana. *Arte Veneta: Rivista Di Storia Dell'arte* 59, 84–97.

Ekman, P. 1992. An argument for basic emotions. *Cognition and Emotion* 6, 169–200.

Elias, J. and Fiore, S. M. 2012. Commenary on the coordinates of coordination and collaboration. In Salas, Eduardo, Stephen M. Fiore and Michael P. Letsky (eds) *Theories of Team Cognition: Cross-disciplinary Perspectives* (571–596). London: Routledge.

Emmons, R. A. 2005. Emotion and religion. In R. F. Paloutzian and C. L. Park (eds) *Handbook of the Psychology of Religion and Spirituality* (235–252). New York: Guilford Press.

Fellbaum, C. (ed.) 1998. *WordNet: An Electronic Lexical Database*. Cambridge, MA: MIT Press.

Flanagan, O. J. 1992. *Consciousness Reconsidered*. Cambridge, MA: MIT Press.

Flew, A. G. N. 1966. *God and Philosophy*. London: Hutchinson

Fuller, R. C. 2009. *Wonder: From Emotion to Spirituality*. The University of North Carolina Press.

Gadamer, H.-G. 1989. *Truth and Method*. 2nd ed. London: Sheed and Ward.

Galileo G. 1632. *Dialogue Concerning the Two Chief World Systems: Ptolemaic & Copernican*. Berkeley: University of California Press, 1953.

Gallagher, S. 1992. *Hermeneutics and Education.* Albany: State University of New York Press.
Gallagher, S. 2003. Phenomenology and experimental design. *Journal of Consciousness Studies* 10 (9–10), 85–99.
Gallagher, S. 2004. Hermeneutics and the cognitive sciences. *Journal of Consciousness Studies* 11 (10–11), 162–174.
Gallagher, S. 2005. *How the Body Shapes the Mind.* Oxford: Oxford University Press.
Gallagher, S. 2007. Neurophilosophy and neurophenomenology. In L. Embree and T. Nenon (eds) *Phenomenology 2005* Vol. 5. (293–316). Bucharest: Zeta Press.
Gallagher, S. 2008. Direct perception in the intersubjective context. *Consciousness and Cognition* 17(2), 535–543.
Gallagher, S. 2009. Two problems of intersubjectivity. *Journal of Consciousness Studies* 16, 298–308.
Gallagher, S. 2013. The socially extended mind. *Cognitive Systems Research* 25–26, 4–12.
Gallagher, S. (in press). Do we (or our brains) actively represent *or* enactively engage with the world? In A. K. Engel, K. Friston, and D. Kragic (eds.), *Where's the Action? The Pragmatic Turn in Cognitive Science.* Cambridge, MA: MIT Press.
Gallagher, S. and Aguda, B. 2015. The embodied phenomenology of phenomenology. *Journal of Consciousness Studies* 22(3–4): 93–107.
Gallagher, S. and Bower, M. 2014. Making enactivism even more embodied? *AVANT/Trends in Interdisciplinary Studies* (Poland) 5(2), 232–247.
Gallagher, S. and Brøsted Sørensen, J. 2006. Experimenting with phenomenology. *Consciousness and Cognition* 15(1), 119–134.
Gallagher, S., Reinerman, L., Sollins, B. and Janz, B. 2014. Using a simulated environment to investigate experiences reported during space travel. *Theoretical Issues in Ergonomic Sciences* 15(4), 376–394.
Gallagher, S. and Varela, F. J. 2003. Redrawing the map and resetting the time: phenomenology and the cognitive sciences. *Canadian Journal of Philosophy. Supplementary Volume* 29, 93–132.
Gallagher, S. and Varga, S. 2014. Social constraints on the direct perception of emotions and intentions. *Topoi* 33, 185–199.
Gallagher, S. and Zahavi, D. 2012. *The Phenomenological Mind.* London: Routledge.
Gallese, V. 2014. Bodily selves in relation: embodied simulation as second-person perspective on intersubjectivity. *Philosophical Transactions of the Royal Society B: Biological Sciences*, 369(1644), DOI: 10.1098/rstb.2013.0177.
Galsion, P. and Jones, C. A. (eds) 1998. *Picturing Science, Producing Art.* New York/ London: Routledge.
Geppert, A. C. T. (ed.) 2012. *Imagining Outer Space.* Basingstoke: Palgrave Macmillan.
Gevins, A., Smith, M. E., McEvoy, L. and Yu, D. 1997. High-resolution EEG mapping of cortical activation related to working memory: effects of task difficulty, type of processing, and practice. *Cerebral Cortex* 7(4), 374–385.
Gevins, A. S., Yeager, C. L., Diamond, S. L., Spire, J., Zeitlin, G. M., and Gevins, A. H. 1975. Automated analysis of the electrical activity of the human brain (EEG): a progress report. *Proceedings of the IEEE* 63(10), 1382–1399.

Gibson, J. J. 1977. The theory of affordances. In J.J. Gieseking, W. Mangold, C. Katz, S. Low, S. Saegert (eds) *The People, Place, and Space Reader* (56–60). Hilldale, NJ: Lawrence Erlbaum.

Gibson, J. J. 2013. *The Ecological Approach to Visual Perception*. New York: Psychology Press.

Goldman, A. I. 2012. A moderate approach to embodied cognitive science. *Review of Philosophy and Psychology* 3(1), 71–88.

Graesser, A. C., McNamara, D. S., Louwerse, M. M. and Cai, Z. 2004. Coh-Metrix: analysis of text on cohesion and language. *Behavior Research Methods, Instruments, and Computers* 36, 193–202.

Harris, C. S. 1965. Perceptual adaptation to inverted, reversed, and displaced vision. *Psychological Review* 72, 419–444.

Harris, L. R., Jenkin, M., Jenkin, H., Dyde, R., Zacher, J. and Allison, R. S. 2010. The unassisted visual system on earth and in space. *J Vestibular Research* 20, 25–30.

Hartl, G. and Sicka, C. 2005 Komposition oder Abbild? Die Darstellung des Nachhimmels in Adam Elsheimers Flucht nach Ägypten – eine naturwissenschaftlich-kritische Betrachtung. In R. Baumstark, M. Dekiert and W. Augustyn (eds) *Von Neuen Sternen: Adam Elsheimers* Flucht Nach Ägypten (107–125). München [u.a.]: Pinakothek-DuMont.

Haugeland, J. 1985. *Artificial Intelligence: The Very Idea*. Cambridge, MA: MIT Press.

Heidegger, M. 1992. *Parmenides*. Trans. A. Schuwer and R. Rojcewicz. Bloomington: Indiana University Press.

Hügel, F. von. 1908. *The Mystical Element of Religion: As Studied in Saint Catherine of Genoa and Her Friends*. 2 vols. London: J. M. Dent & Co.

Hurley, S. 1998. *Consciousness in Action*. Cambridge, MA: Harvard University Press.

Husserl, E. 1970. *The Crisis of the European Sciences and Transcendental Phenomenology*. Evanston, IN: Northwestern University Press.

Husserl, E. 2012. *Ideas: General Introduction to Pure Phenomenology*. London: Routledge.

Idler, E. L., Musick, M. A., Ellison, C. G., George, L. K., Krause, N. and Williams, D. R. 2003. Measuring multiple dimensions of religion and spirituality for health research. *Research on Aging* 25, 327–365.

Ishihara, S. 2010. *Ishihara's Tests for Colour Deficiency*. 14 Plate Book Concise Edition (1st ed.) Graham-Field.

James, W. 1890. *The Principles of Psychology. 2 Vols*. New York: Dover 1950.

Keltner, D. and Haidt, J. 2003. Approaching awe, a moral, spiritual, and aesthetic emotion. *Cognition & Emotion* 17(2), 297.

Kennedy, R. S., Lane, N. E., Berbaum, K. S. and Lilienthal, M. G. 1993. Simulator sickness questionnaire: an enhanced method for quantifying simulator sickness. *International Journal of Aviation Psychology* 3(3), 203–220.

Kingwell, M. 2000. Husserl's sense of wonder. *The Philosophical Forum* 31(1), 85–107.

Kozin, A. V. 2007. Iconic wonder: Pavel Florensky's phenomenology of the face. *Studies in Eastern European Thought* (59), 293–308.

Kramer, L. A., Sargsyan, A. E., Hasan, K. M., Polk, J. D. and Hamilton, D. R. 2012. Orbital and intracranial effects of microgravity: findings at 3-T MR Imaging. *Radiology* 263(3), 819–827.

Le Van Quyen, M. and Petitmengin, C. 2002. Neuronal dynamics and conscious experience: an example of reciprocal causation before epileptic seizures. *Phenomenology and the Cognitive Sciences* 1(2), 169–180. doi:10.1023/A:1020364003336

López-Alegría, M. 2014. Looking down at the earth. In M. Najjar (ed.) *Outer Space* (68–90). Berlin Distanz Verlag.

Lovelock, J. 2000. *Gaia: A New Look at Life on Earth.* Oxford: Oxford University Press.

Lutz, A. 2002. Toward a neurophenomenology as an account of generative passages: a first empirical case study. *Phenomenology and the Cognitive Sciences* 1, 133–167.

Lutz, A., Lachaux J. -P., Martinerie J. and Varela F. J. 2002. Guiding the study of brain dynamics by using first-person data: synchrony patterns correlate with ongoing conscious states during a simple visual task. Proceedings of the National Academy of Sciences 99, 1586–1591

Lutz, A. and Thompson, E. 2003. Neurophenomenology integrating subjective experience and brain dynamics in the neuroscience of consciousness. *Journal of Consciousness Studies* 10(9–10), 31–52.

Mader, T. H., Gibson, C. R., Pass, A. F., Kramer, L. A., Lee, A. G., Fogarty, J., Tarver, W. J. et al. 2011. Optic disc edema, globe flattening, choroidal folds, and hyperopic shifts observed in astronauts after long-duration space flight. *Ophthalmology* 118(10), 2058–2069.

Magnus, A. 1988. *Super Dionysii Mysticam Theologiam.* Translated in *Albert and Thomas: Selected Writings*, S. Tugwell (trans.). New York: Paulist Press.

Makarchouk, N. E., Maksimovich, K. Y., Kravchenko, V. I., and Kryzhanovskii, S. A. 2011. Modifications of EEG activity related to perception of emotionally colored, erotic, and neutral pictures in women during different phases of the ovulatory (menstrual) cycle. *Neurophysiology* 42(5), 362–370.

Mallis, M. M. and DeRoshia, C. W. 2005. Circadian rhythms, sleep, and performance in space. *Aviation, Space, and Environmental Medicine*, 76(Supplement 1), B94–B107.

Masters, K. S. 2013. Brief multidimensional measure of religiousness/spirituality (BMMRS). In *Encyclopedia of Behavioral Medicine* (267–269). New York: Springer.

Masters, K. S., Carey, K. B., Maisto, S. A., Caldweel, P. E., Wolfe, T. V., Hackney, H. L., France, C. R. and Himawan, L. 2009. Psychometric examination of the brief multidimensional measure of religiousness/spirituality among college students. *International Journal for the Psychology of Religion* 19, 106–120.

Mayer, H. 2011. A short chronology of spaceflight. In C. Brünner and A. Soucek (eds) *Outer Space in Society, Politics and Law* (20–29). Vienna and New York.

McClain, D. L. 2009. Evidence of the properties of an ambiguity tolerance measure: the Multiple Stimulus Types Ambiguity Tolerance Scale-II (MSTAT-II). *Psychological Reports* 105(3), 975–988.

McDonald, J. D. 2008. Measuring personality constructs: the advantages and disadvantages of self-reports, informant reports and behavioural assessments. *Enquire* 1, 1–18.

McGivern, P. 2008. Reductive levels and multi-scale structure. *Synthese* 165(1), 53–75.

McIndoo, J. M. 1914. Instinct as Related to Education. PhD Dissertation, Clark University. Worcester, MA.

Mclain, D. L. 1993. The Mstat-I: a new measure of an individual's tolerance for ambiguity. *Educational and Psychological Measurement* 53, 183–189.

Melzack, R. and Wall, P. D. 1967. Pain mechanisms: a new theory. *Survey Instinct as Rrelated to Educationof Anesthesiology* 11(2), 89–90.

Merleau-Ponty, M. 1964. *The Structure of Behavior.* Trans. A. L. Fisher. Boston: Beacon Press.

Merleau-Ponty, M. 1968. *The Visible and the Invisible.* Trans. A. Lingis. Evanston: Northwestern University Press.

Merleau-Ponty, M. 2012. *Phenomenology of Perception.* Trans. D. A. Landes. London: Routledge.

Metzinger, T. 2004. *Being No One: The Self-Model Theory of Subjectivity.* Cambridge, MA: MIT Press.

Miller, G. A, Beckwith, R., Fellbaum, C., Gross, D. and Miller, K. 1990. Five papers on WordNet. Report No. 43. Princeton, NJ: Cognitive Science Laboratory, Princeton University.

Mitchell, S. D. 2002. Integrative pluralism. *Biology and Philosophy* 17(1), 55–70.

Mühling, M. 2008. *The Space that Time Forgot: Angela Bulloch, Städtische Galerie Im Lenbachhaus Und Kunstbau, München.* Köln: König.

Nagel, T. 1974. What is it like to be a bat? *The Philosophical Review* 83(4), 435–450.

NASA *Science News.* 2001. Mixed up in space. At http://science.nasa.gov/science-news/science-at-nasa/2001/ast07aug_1/.

Nicogossian, A. E. and Parker Jr, J. F. 1982. *Space Physiology and Medicine* (No. N-8325349; NASA-SP-447). National Aeronautics and Space Administration, Washington, DC (USA).

Neger, T. and Soucek, A. 2011. Space faring: a short overview of the present situation (2.4.1 The space race after WWII). In C. Brünner and A. Soucek (eds) *Outer Space in Society, Politics and Law* (157–159). Vienna and New York.

Newberg, A. B. 1994. Changes in the central nervous system and their clinical correlates during long-term spaceflight. *Aviation, Space, and Environmental Medicine* 65(6), 562–572.

Newberg, A. B. and Newberg, S. K. 2005. The neuropsychology of religious and spiritual experience. In R. F. Paloutzian and C. L. Park (eds) *Handbook of the Psychology of Religion and Spirituality* (199–215). New York: Guilford Press.

Newell, C. L. 2013. The strange case of Dr. von Braun and Mr. Disney: Frontierland, Tomorrowland, and America's final frontier. *Journal of Religion and Popular Culture* 3, 416.

Noë, A. 2004. *Action in Perception.* Cambridge, MA: MIT Press.

Novak, J. D. and Cañas, A. J. 2006. The theory underlying concept maps and how to construct them. *Florida Institute for Human and Machine Cognition, 1.* Technical Report IHMC CmapTools 2006–2001

Oppenheim, P. and Putnam, H. H. 1958. Unity of science as a working hypothesis. *Minnesota Studies in the Philosophy of Science* 2, 3–36.

O'Regan, K. and Noë, A. 2001. A sensorimotor account of vision and visual consciousness. *Behavioral and Brain Sciences* 23, 939–973.

Osborne, L. 1997. *The New Vision of the Earth.* Cocoa, FL: Theater 360, Brevard Community College.

Paulhus, D. L. and Vazire, S. 2009. The self-report method. In R.W. Robins, R. C. Fraley and R. F. Krueger (eds) *Handbook of Research Methods in Personality Psychology* (224–239). New York: Guilford Press.

Paus, T. and Zatorre, R. J. 1997. Time-related changes in neural systems underlying attention and arousal during the performance of an auditory vigilance task. *Journal of Cognitive Neuroscience* 9(3), 392.

Penrose, R. 1999. *The Emperor's New Mind: Concerning Computers, Minds, and the Laws of Physics*. Oxford: Oxford University Press.

Pepper, K. 2014. Do sensorimotor dynamics extend the conscious mind? *Adaptive Behavior* 22(2), 99–108.

Petitmengin C. 2006. Describing one's subjective experience in the second person: an interview method for the science of consciousness. *Phenomenology and the Cognitive Sciences* 5, 229–269

Petitmengin, C. 2010. A neurophenomenological study of epileptic seizure anticipation. In S. Gallagher and D. Schmicking (eds) *Handbook of Phenomenology and Cognitive Science* (471–499). Springer Netherlands.

Petitot J., Varela, F., Roy, J-M. and Pachoud, B. 1999. *Naturalizing Phenomenology: Issues in Contemporary Phenomenology and Cognitive Science*. Stanford, CA: Stanford University Press

Pfurtscheller, G. and Lopes da Silva, F. H. 1999. Event-related EEG/MEG synchronization and desynchronization: basic principles. *Clinical Neurophysiology* 110(11), 1842–1857.

Pogosyan, A., Gaynor, L. D., Eusebio, A. and Brown, P. 2009. Boosting cortical activity at beta-band frequencies slows movement in humans. *Current Biology* 19(19), 1637–1641. doi:10.1016/j.cub.2009.07.074

Prinzel, L. J., Freeman, F. G., Scerbo, M. W., Mikulka, P. J. and Pope, A. T. 2000. A closed-loop system for examining psychophysiological measures for adaptive task allocation. *The International Journal of Aviation Psychology* 10(4), 393–410.

Reddy, V. 2008. *How Infants Know Minds*. Cambridge, MA: Harvard University Press.

Reinerman-Jones, L. E., Matthews, G., Langheim, L. K. and Warm, J. S. 2010. Selection for vigilance assignments: a review and proposed new directions. *Theoretical Issues in Ergonomics Science* 1–24.

Reinerman-Jones, L. E., Sollins, B., Gallagher, S. and Janz, B. 2013. Neurophenomenology: an integrated approach to exploring awe and wonder. *South African Journal of Philosophy* 32 (4), 295–309.

Rieth, A. 1953. *Der Blitz in der bildenden Kunst*, München: Heimeran

Rock, I. and Harris, C. S. 1967. Vision and touch. *Scientific American*, 216(5), 96–104.

Roll, J-P. and Roll, R. 1988. From eye to foot: a proprioceptive chain involved in postural control. In G. Amblard, A. Berthoz, and F. Clarac (eds) *Posture and Gait: Development, Adaptation, and Modulation* (155–164). Amsterdam: Excerpta Medica.

Rowlands, M. 2010. *The New Science of the Mind*. Cambridge, MA: MIT Press.

Schacter, D. L. 1977. EEG theta waves and psychological phenomena: a review and analysis. *Biological Psychology* 5(1), 47–82.

Shapiro, L. A. 2004. *The Mind Incarnate*. Cambridge, MA: MIT Press.

Sherman, W. R. and Craig, A. B. 2002. *Understanding Virtual Reality: Interface, Application, and Design*, 1st Edn. San Francisco, CA: Morgan Kaufmann.Sherman W. R., Craig A. B. (2002). *Understanding Virtual Reality: Interface, Application, and Design*, 1st Edn. San Francisco, CA: Morgan Kaufmann

Shiota, M. N., Keltner, D. and Mossman, A. 2007. The nature of awe: elicitors, appraisals, and effects on self-concept. *Cognition & Emotion* 21(5), 944–963
Smith, M. E., Gevins, A., Brown, H., Karnik, A. and Du, R. 2001. Monitoring task loading with multivariate EEG measures during complex forms of human-computer interaction. *Human Factors* 43(3), 366–380.
Spinoza, de B. 1970. *Ethics*. J. Gutmann (ed.) New York: Simon and Schuster.
Steinbock, A. 2007. *Phenomenology and Mysticism: The Verticality of Religious Experience*. Bloomington: Indiana University Press.
Straus, E. 1966. *Philosophical Psychology*. New York: Basic Books.
Strollo, F. 1999. Hormonal changes in humans during spaceflight. *Adv Space Biol Med* 7(2), 99–129.
Sutton, J., Harris, C. B., Keil, P. G. and Barnier, A. J. 2010. The psychology of memory, extended cognition, and socially distributed remembering. *Phenomenology and the Cognitive Sciences* 9(4), 521–560.
Taddei-Ferretti, C. and Musio, C. 1999. The neural net of Hydra and the modulation of its periodic activity. In J. Mira and J. V. Sánchez-Andrés (eds) *Foundations and Tools for Neural Modeling* 1606 (123–137). Berlin: Springer.
Task H. L. and Genco L. V. 1987. Effects of short-term space flight on several visual functions. In B. W. Bungo, T. M. Bagian, M. A. Bowman and B. M. Levitan (eds), Results of the bye sciences DSOs conducted aboard the space shuttle 1981–1986. NASA Technical Memorandum 58280, p. 173
Taylor, C. 1985. *Human Agency and Language*. Cambridge: Cambridge University Press.
Tellegen, A. and Atkinson, G. 1974. Openness to absorbing and self-altering experiences ('absorption'), a trait related to hypnotic susceptibility. *Journal of Abnormal Psychology* 83, 268–277. doi:10.1037/h0036681
Thompson, E. 2007. *Mind in Life: Biology, Phenomenology and the Sciences of Mind*, Cambridge, MA: Harvard University Press.
Thompson E., Lutz A. and Cosmelli D. 2005. Neurophenomenology: an introduction for neurophilosophers. In A. Brook and K. Akins (eds) *Cognition and the Brain: The Philosophy and Neuroscience Movement* (40–97). New York, NY: Cambridge University Press.
Thompson, E. and Stapleton, M. 2009. Making sense of sense-making: reflections on enactive and extended mind theories. *Topoi* 28(1), 23–30.
Thompson, E. and Varela, F. 2001. Radical embodiment: neural dynamics and consciousness. *Trends in Cognitive Sciences* 5(10), 418–425.
Tourangeau, R. 2000. Remembering what happened: memory errors and survey reports. In A. A. Stone et al. (eds) *The Science of Self-report: Implications for Research and Practice* (29–48). New Jersey: Lawrence Erlbaum.
Trevarthen, C. B. 1979. Communication and cooperation in early infancy: a description of primary intersubjectivity. In M. Bullowa (ed.) *Before Speech* (321–348). Cambridge: Cambridge University Press.
Trevarthen, C. and Hubley, P. 1978. Secondary intersubjectivity: confidence, confiding and acts of meaning in the first year. In A. Lock (ed.) *Action, Gesture and Symbol: The Emergence of Language* (183–229). London: Academic Press.
Tulving, E., Markowitsch, H. J., Craik, F. I., Habib, R. and Houle, S. 1996. Novelty and familiarity activations in PET studies of memory encoding and retrieval. *Cerebral Cortex* 6(1), 71–79. doi:10.1093/cercor/6.1.71

Turing, A. M. 1950. Computing machinery and intelligence. *Mind* 59(236), 433–460.

Underhill, E. 1912. *Mysticism: A Study in the Nature and Development of Man's Spiritual Consciousness*. 4th Ed. London: Methuen & Co. Ltd.

van Riel, R. and Van Gulick, R. 2014. Scientific reduction. In E. N. Zalta (ed.) *The Stanford Encyclopedia of Philosophy*. At: http://plato.stanford.edu/archives/sum2014/ entries/scientific-reduction/.

Varela F. J. 1996. Neurophenomenology: a methodological remedy for the hard problem. *Journal of Consciousness Studies* 3, 330–349

Varela, F. J., Thompson, E. and Rosch, E. 1991. *The Embodied Mind: Cognitive Science and Human Experience*. Cambridge: MIT Press.

Villard, E., Garcia-Moreno, F. T., Peter, N. and Clément, G. 2005. Geometric visual illusions in microgravity during parabolic flight. *Neuroreport* 16(12), 1395–1398.

Wallace, B. A. 2007. *Where Buddhism and Neuroscience Converge*. New York: Columbia University Press.

Wapner, S. and Werner, H. 1965. An experimental approach to body perception from the organismic developmental point of view. In S. Wapner and H. Werner (eds) *The Body Percept*. New York: Random House.

Ward, B. 1968. *Spaceship Earth*. New York: Columbia University Press.

Weems, S. A., Zaidel, E., Berman, S. and Mandelkern, M. A. 2004. Asymmetry in alpha power predicts accuracy of hemispheric lexical decision. *Clinical neurophysiology* 115(7), 1575–1582. doi:10.1016/j.clinph.2004.02.020

Weinstein, D. 2002. Captain video: television's first fantastic voyage. *Journal of Popular Film and Television* 3, 148.

White, F. 1987. *The Overview Effect: Space Exploration and Human Evolution*. Reston, VA: The American Institute of Aeronautics and Astronautics.

Wilkinson, M. J., Lulla, K., Bowersox, K. D., Rominger, K., Thornton, K., Coleman, C., Lopez-Alegria, M., Leslie, F. and Sacco, A. 1998. Space shuttle Columbia's earth observations photography (October 20–November 5, 1995): A mission report. *Geocarto International* 13(1), 83–97.

Wubbels, P., Nishimura, E., Rapoport, E., Darling, B., Proffitt, D., Downs, T. and Downs III, J. H. 2007. Exploring calibration techniques for functional near-infrared imaging (fNIR) controlled brain-computer interfaces. In D. Schmorrow (ed.), *Foundations of Augmented Cognition* (23–29). Berlin: Springer.

Zahavi, D. 2010. Naturalized phenomenology. In S. Gallagher and D. Schmicking (eds) *Handbook of Phenomenology and Cognitive Science* (2–19). Dordrecht: Springer.

Zajac, F. E. 1993. Muscle coordination of movement: a perspective. *Journal of Biomechanics* 26(suppl. 1), 109–124.

Zwart, S. R., Gibson, C. R., Mader, T. H., Ericson, K., Ploutz-Snyder, R., Heer, M. and Smith, S. M. 2012. Vision changes after spaceflight are related to alterations in folate–and vitamin B-12–dependent one-carbon metabolism. *The Journal of Nutrition* 142(3), 427–431.

Index

Printed and bound by CPI Group (UK) Ltd, Croydon, CR0 4YY